BLUEGRASS

BOURBON BARONS

BLUEGRASS

BOURBON BARONS

BRYAN S. BUSH

AMERICAN PALATE

Published by American Palate
A Division of The History Press
Charleston, SC
www.historypress.com

First published 2021

Manufactured in the United States

ISBN 9781467150132

Library of Congress Control Number: 2021937164

CONTENTS

ACKNOWLEDGEMENTS

Special thanks must go to the Louisville Free Public Library. Without its research tools that gave me access to several different newspapers, I would have never have been able to write this book.

I would also like to thank Cave Hill Cemetery in Louisville, Kentucky, for its support throughout the years, in particular Michael Higgs, coordinator of the Cave Hill Heritage Foundation, and Alex Lueken. Cave Hill Cemetery is Louisville's premier cemetery. The purpose of the Cave Hill Heritage Foundation is to secure funding for the long-term preservation of this unique cemetery. Specifically, the mission of the Cave Hill Heritage Foundation is threefold:

- To Restore the Historical Monuments and Buildings: Cave Hill Cemetery is known for its exquisite collection of monumental art, many examples of which are more than 150 years old. Additionally, the property includes a variety of historic structures, from the administration office to the board room to the three-and-a-half-mile-long brick and stone wall that encircles the perimeter. The resources provided through the foundation will allow cemetery management to be proactive both in addressing specific, long-term conservation projects and in responding to critical situations that require immediate intervention and stabilization.

- TO PRESERVE THE ARBORETUM SETTING: On average, fifty trees are removed each year at Cave Hill Cemetery due to disease, old age or weather damage, and more than one hundred plates are replaced. The Cave Hill Heritage Foundation will help provide resources for the unexpected, as well as help preserve and sustain the beauty of the cemetery's arboretum setting. It will provide for the removal and the replacement of specimen trees and shrubs and ensure that future generations will have access to this amazing green space in the heart of Louisville.

- TO PROVIDE COMMUNITY EDUCATION AND AWARENESS: The history of Cave Hill Cemetery is inextricably tied to the history of Louisville. Through its landscape and monuments, the cemetery tells the story of this beloved city by the river and its remarkable citizenry. The Cave Hill Heritage Foundation will help provide resources to produce educational materials, expand public awareness and develop special events and programs for schoolchildren and the community at large. It will also ensure that the critical task of recording and archiving the history of the cemetery will continue.

I would also like to thank my parents, Carol and Gene Bush, and stepmom, JoAnn Bush, for their support. I would also like to thank my friend Ryan Reynolds for support and inspiration.

INTRODUCTION

The American Industrial Revolution changed the bourbon industry during the Gilded Age and transformed the spirit from small-batch production to the bourbon that Americans know today. Whiskey bourbon became the leading industrial product in the state. There are several areas in Kentucky that claim to be the site of the first manufacture of whiskey. Mason County claims that the first Kentucky distillery was located a few miles from the present town of Maysville. Nelson County also claims to be the first location of a distillery. Mercer County claims that the first distillery was built in Harrodsburg, which was the earliest permanent settlement in Kentucky.

General James Wilkinson was the owner and operator of the distillery in Harrodsburg, but he did not arrive in Kentucky until 1784. In 1783, Evan Williams built and operated a still at the corner of Fifth and Water Streets in Louisville at the Falls of the Ohio. He used the abundant supply of corn to make his whiskey. Williams was Kentucky's first commercial distiller. Williams was not only a distiller but also a member of the Louisville Board of Trustees. Before he became a member of the board, a rule had been adopted forbidding liquor during the meetings, under penalty of censure and the fine of six shillings; if the law were broken a second time, the board would seize the liquor. Williams did not know about the rule and brought a bottle of his own whiskey to the meeting for the refreshment of his fellow members. The members seized his liquor, but without censure. At the end of the meeting, Williams left with an empty bottle. At the next meeting,

Williams brought another bottle, but Will Johnson, the county clerk, declared that Williams ought to be expelled for making and offering to the board his whiskey. The clerk was also surveyor of the port and was accustomed to the use of imported wines and liquors. In his defense, Williams insisted that the clerk had the taste of an aristocrat, and the board agreed with him—again Williams left with an empty bottle, without censure. Williams continued to make his whiskey without license until 1788, when he was forced to pay for his license and continued to run his distillery.[1] In the 1780s, James Spears, a distiller from Paris, Kentucky, first used the term *bourbon*.

In 1787, a great boost in the manufacture of whiskey in Kentucky arose when a colony of immigrants from Maryland arrived in Kentucky. They were known as the "Catholic League." Sixty families came down the Ohio River in flatboats and settled in Nelson, Washington and Marion Counties. They brought with them the knowledge of distilling and established early distilleries along the Salt River. After manufacturing the whiskey, they would send their product down the river to the settlements and the cities of the Ohio and Mississippi Rivers. Nelson County possessed an abundant supply of limestone water. Nelson County was and still remains the principal center for distilleries in Kentucky and made the finest whiskies in the state.[2]

Elijah Craig was credited with being the first distiller to age his whiskey in charred oak barrels. No one knows how Elijah began charring his barrels. Once he began using charred oak, the whiskey changed into an amber color, with a distinctive flavor that makes what we now know today know as bourbon. Many consider Elijah Craig as the father of bourbon.[3]

In 1791, an excise law was passed by Congress for the purpose of raising federal revenue. Sixty distillers in Pennsylvania refused the pay the tax, and in July 1794, federal marshal David Lenox began the process of collecting the tax. His guide was John Neville of Allegheny County. As many as seven hundred men approached Neville's home and demanded his surrender. Ten soldiers protected the property. The mob demanded the soldiers surrender. The soldiers and mob fired on each other, and soon the soldiers surrendered; Neville's Bower Hill estate was burned to the ground. President George Washington told the insurgents to disperse and warned the people not to resist the federal government. The militia was called out and directed to quell the rebellion. The insurgents backed down, and General Harry Lee issued a proclamation granting amnesty to all who had submitted to the laws; the insurgents had to take an oath of allegiance. The insurgence was known as the Whiskey Rebellion of Pennsylvania.

With the law passed, many of the distillers in Pennsylvania moved to Kentucky, settling in Mason County. The government of Kentucky did not interfere with the distillers, and they were able to convert their corn into whiskey without a tax. Whiskey production occupied the pioneers of Kentucky. Whiskey became a form of exchange and a standard of value. The product was in universal demand and did not deteriorate; in fact, it actually improved with age. The product could be easily transported and could be disposed of in any town or cities along the Ohio River or Mississippi River. Settlements of accounts were made in whiskey, and many professional men received their wages in whiskey.[4] In 1794, Alexander Anderson of Philadelphia introduced the perpetual steam still. By 1801, whiskey and tobacco had replaced flour as the principal export crop from Kentucky's interior, with 56,000 gallons traveling down the Ohio River and passing through the Louisville Customs House. By 1810, 250,000 gallons were passing through the customs house, and by 1822, 2,250,000 gallons of whiskey were being transported down the Ohio River.[5] By 1815, the steam-powered perpetual still was being used as far west as Kentucky, and by 1818, it had replaced the copper still that George Washington and Evan Williams would have used in their whiskey production. By 1830, Aeneas Coffey had invented the column still, an advanced variant of the Alexander Anderson perpetual still. Coffey's still could produce 3,000 gallons of raw whiskey in an hour.[6]

Scotchman James Crow studied medicine and chemistry at Edinburg University. In 1823, he immigrated to Kentucky and opened his own distillery. He owned Glenn's Creek Distillery and later the Old Oscar Pepper Distillery. He applied the scientific method to distilling. He used saccharimeters to gauge the sugar content of the mash and thermometers to regulate the mash's temperature and took careful measurements of acidity levels throughout the process. His contribution to the history of the sour mash process for regulatory batch consistency came about when he added a small quantity of leftover mash (the mixture of grain, malt, water and yeast) from the previous batch. By doing so, he was able to regulate the pH levels in each run of the still, enabling consistency on a large scale.[7]

With the invention of bigger and more efficient stills and James Crow's development of palatable bourbon, manufacturers could produce a quality product on a large scale; nonetheless, whiskey bourbon began to decline, and by 1846, there were no distilleries in Louisville. In 1821, the Hope Distillery turned out one thousand gallons of whiskey, but the company was a failure because it ran out of people to drink the product. Steam power, modern buildings, advanced machinery and high-volume stills helped with production, but once made, the whiskey bourbon could only be delivered by wagon or steam-powered boats, which could take months to deliver the product to the East Coast or New Orleans.[8]

After the Civil War, the manufacture of whiskey made a comeback. Railroads changed how bourbon was delivered. By 1865, there were 35,000 miles of track; by 1870, 53,000 miles of track; by 1880, 93,000 miles of track; and by 1890 164,000 miles of track. When the Louisville and Nashville Railroad connected with other railroads, the trains could reach the East, West, South and North, delivering bourbon over the entire county and even overseas on steam-powered ocean liners. The American Industrial Revolution changed the farm, bringing about modern equipment such as Avery's plows and artificial fertilizers. Corn production in 1860 was 838.8 million bushels per year, but by 1900, corn production had risen to 2.7 billion bushels per year. More corn meant more whiskey and hence more bourbon.[9] But during the Gilded Age (1870–1900), the major corporations took over the small stills and turned the business of bourbon into a multimillion-dollar industry. The building of bonded warehouses, the knowledge of intricate laws and rulings required and the great expense essential to meet all the demands of the government, together with the capital required to meet the taxes as they matured, had the effect of gradually replacing the farmer who made whiskey on a small scale and putting the manufacturing of the product into the hands of corporations and individuals with strong financial resources. The bourbon industry was almost ruined by deceptive practices such as adding flavoring or colors to bourbon to mask inferior quality. The master distillers in Kentucky took back their product and demanded regulations. Batches were tested, and bourbon was bottled with bonded labels to make sure the product was not tampered with; government regulations ensured that what was in the bottle actually was bourbon.

In 1880, Kentucky saw 15,011,279 gallons of whiskey made in the state, but by 1895, the gallons of whiskey produced had increased to 22,814,950. Kentucky distillers spent millions of dollars per year in the purchase of grain, fuel, barrels and machinery, as well as for labor, insurance and transportation. In 1895, there were nineteen distillers in Louisville, plus handling the outputs of Nelson, Davies, Anderson and other counties. Several of the distilleries had the warehouse capacity for more than 100,000 barrels of whiskey. The production of and trade in fine bourbons was one of the greatest industries in Kentucky and took up a large amount of capital in Louisville. Louisville was the collection center and the site of one hundred registered grain distilleries. The producing capacity was 82,000 gallons per day. The gross product during the five years ending in June 1887 was 35,000 gallons, with internal revenue taxes amounting to $29,154,319.[10] By 1904, 436,013 barrels were being shipped from Louisville, which converts to 21 million gallons of whiskey bourbon. Louisville handled one-sixth of the total consumption of whiskey of the United States and was the largest market for fine whiskey in the United States. The value of the shipments was $36,991,040. The total production of whiskey bourbon in the state of Kentucky for 1904 was 23,070,162 gallons, of which 11,398,394 gallons were produced in Louisville. By the turn of the century, Louisville was leading the world in whiskey bourbon.[11]

Here are just some of the distillers that have been located in Louisville: J.M. Mattingly & Sons, located on High Avenue in Portland, with its office on 205 West Main Street; Block, Franck & Company, located on 205 West Main; J.B. Wathen & Brothers Company, located on 141 West Main Street; the Anderson & Nelson Distilleries Company, located on 116 East Main Street; Applegate & Sons, located on 122 East Main Street; the Parkland Distillery Company, located on 126 East Main Street, Hollenbach & Vetter; owner of the Glencoe Distillery, located on 234 Second Street; the Ashton Distillery Company, located on 120 East Main; John Roach, maker of "Old Times," located on 104 East Main Street; J.H. Cutter, maker of Old Bourbon; C.P. Moorman & Company, located at 104 East Main Street; the Marion County Distillery, located on 31st Street and Rudd Avenue; and J.M. Atherton Company, located on 125 West Main Street.

Unfortunately, with the passing of the Prohibition Act in 1920, all the bourbon companies in Louisville had to liquidate. Only six distilleries were allowed to make medicinal bourbon during the years of Prohibition. Some of the bourbon founders were able to sell their businesses to a conglomerate called the Kentucky Distilleries and Warehouse Company before Prohibition took effect; some knew that Prohibition was coming and sold their businesses. Only recently have Louisville and many other parts of Kentucky made a comeback in the bourbon trade. The old Whiskey Row on Main Street in downtown Louisville has seen the resurgence of Evan Williams, Old Forrester, Angel's Envy, Rabbit Hole Distillery and, as the most recent addition, Mitchner's Fort Nelson Distillery. The Frazier History Museum has refocused its mission and become the first stop on the Kentucky Bourbon Trail.

The purpose of this book is to reveal the stories of the men who made Louisville the bourbon capital of the world before Prohibition. Some of

Old Forester, made by the Brown-Forman Corporation, built a bourbon experience welcome center on the location where Brown-Forman (under George Garvin Brown and George Forman) had its headquarters. The newly restored Old Forester has re-created a "Whiskey Drummer," who at certain times of the day beats a drum as the distillery rolls out Old Forester whiskey barrels onto its trucks as they make their way to the warehouse. Another bourbon experience welcome center on Whiskey Row is Evan Williams, located on 528 West Main Street in Louisville. *Photo by author.*

Whiskey Row, located on West Main Street in downtown Louisville, was once the location for offices of several of the bourbon barons who controlled Kentucky's number one manufactured product. Today, Whiskey Row has been beautifully restored. *Photo by author.*

the men not only owned bourbon companies but also were foundational in Louisville's economic history. Men like James Atherton took an interest in education and improved the schools in Louisville by implementing a board of education that oversaw the school system—Atherton High School was named after him. There was Isaac Bernheim, who founded Bernheim, which is now called the Bernheim Forest and Research Center; Samuel Grabfelder, who helped found Jewish Hospital; and John Douglas, who built a racetrack for the trotter racing industry and was known as the "Prince of Sports." Some of the bourbon leaders of Louisville served in the Civil War. With such enormous wealth came the temptation for fraud, and this led several bourbon leaders to become involved in some of Kentucky's most famous fraud cases. Each owner, producer or operator has his own unique story. Let us step back in time, revisit the bourbon barons of Kentucky and learn how they laid the groundwork for the success of bourbon today.

CHAPTER 1

COLONEL JOHN JAMES DOUGLAS

Yosemite, Jim Douglas, Douglas Malt, Eagle Elk, Glynn Valley, Winetrop Club, Carlton, Salvator and Meleager

John J. Douglas was born on December 4, 1840, in Louisville, Kentucky. He was the son of James W. Douglas, a miller, and Sarah Janes Douglas. He began his career as a teamster. When he was a teenager, he opened a "policy shop" or lottery on Jefferson Street near Third Street. The lottery shop was one of thirty-nine similar local branches of a lottery owned in New York and operated under a charter from the State of Kentucky. After Douglas spent five years in the lottery shop, his abilities in the business impressed the company's owners, and they made him a manager for the entire state.[12]

Tickets for the monthly drawings were sold all over the country, as well in Louisville and Kentucky, and Douglas handled sales on average of $8,000 per day. For twenty years, he continued as head of the state lottery and during that time amassed a huge fortune. When the company's charter expired, John Douglas added a business partner, B.S. Stewart, in the Frankfort lottery, which was operated for the partial benefit of the schools for Frankfort. For ten years, he operated the lottery and increased his wealth.[13]

In the early 1880s, the popularity of the game began to wane, and shortly afterward, a federal statute was passed that prohibited the use of the mail for the lottery to avoid exploitation. Douglas led the legal battle to prevent the suppression of his lucrative business, bringing him nationwide attention. In 1891, the constitution of Kentucky provided that "lotteries and gift enterprises are forbidden, and no privileges shall be granted for such purposes, and none shall be granted for such purposes and none shall be exercised and no schemes for similar purposes shall be allowed. The General Assembly shall enforce this section by proper penalties. All lottery privileges or charters heretofore granted are revoked." The legal cases were carried through the local courts, the Kentucky Court of Appeals, the federal courts and finally the Supreme Court of the United States. On November 29, 1897, in the case *Douglas v. Kentucky*, the courts upheld the ruling that all lottery privileges or charters were to be revoked.[14] The lottery game was outlawed. Douglas accepted defeat in good spirit and publicly announced that he would not operate in violation of the law.[15]

After giving up the lottery business, he was associated with Lum Simons and William "Bill" Bailey, who was the jailer of Jefferson County. They were engaged in a brokerage and "claim shaving" business.[16]

Later, John Douglas became a wholesale liquor dealer and importer of liquor and formed a company called J.J. Douglas Company. He was a rectifier or blender of other company's bourbons to make Yosemite, Jim Douglas, Douglas Malt, Eagle Elk, Glynn Valley, Winetrop Club, Carlton, Salvator and Meleager. His main office was at 224 West Main Street in Louisville.[17] He was also a turf man and bred, owned and raced both pacers and trotting horses. In 1896, Douglas moved from Louisville to Middletown. His home was located on Shelbyville Road, near Anchorage, and it was one of the showplaces of Middletown. The home was purchased from the Finzer estate. Nicholas Finzer was in the tobacco business along with his brothers, and his widow, Agnes Finzer, sold the farm to Douglas. The farm consisted of four hundred acres. A spring on his property was once owned by the first settlers of Middletown. With fresh water from the spring, families began to build their homes on what would become the Douglas property. Douglas always kept the spring open to the public. On his farm, his property was keenly landscaped, with not a weed or stump to be seen.[18]

He called his home Douglas Place but later renamed the stables the Eothen Stock Farm. His home was one of the most famous in the country at the time. He was successful as a breeder of trotters and pacers. The best-known horse foaled on his farm was Heno, which was sold to John E. Madden for $500.[19] Later, the horse was valued in the thousands. McDole was the best trotter Douglas ever owned. Another famous trotter Douglas owned was Joe Johnson.[20]

In September 1895, to coincide with the opening of the Grand Army of the Republic Reunion in Louisville, Douglas personally founded and opened Douglas Park. He thought that trotters and pacing horses should have a premium venue and purchased property located south of

The J.J. Douglas Company advertisement. *From* Wine and Spirit Bulletin *(1903)*.

Louisville in the Beechmont suburbs where Southside Drive joins Second Street. He constructed a one-mile oval that featured bank turns. A large grandstand was also constructed, along with stables and a clubhouse.[21] Not too long after the opening of Douglas Park, Douglas was made a Kentucky colonel by the governor. In December 1905, through Colonel Sum Simons, Douglas disposed of all his horses, both trotters and pacers, at a dispersal sale in Lexington. The richest stake in the world was Pilatus, which sold for $1,800, which is equivalent to $51,362 today. Free Fancy brought $120, Cecil Wilkes brought $210, Eggatine brought $120, Mayne Nutwood brought $300, Tiggaliska brought $400 and Dallantus brought $200.[22] Also in December 1905, Douglas sold Douglas Park to Louis Cella of the Western Jockey Club. The track sold for $100,000. Colonel Simons conducted the sale of the Douglas track. Simon made the trip to St. Louis, where he met Louis Cella and closed the deal for the sale of the racetrack. According to newspaper reports, Simons received $10,000 for his holdings.[23]

The newly formed incorporation for Douglas Park stated that the track would encourage and promote agriculture and the improvement of stock—particularly running, trotting and pacing horses—by giving exhibits or contests of speed and races between horses for premiums, purses or other rewards. The incorporation would also maintain a fairgrounds and racetrack.[24] Louis Cella had 2,490 shares in the new corporation. In 1907, Churchill Downs and Douglas Park formed a holding company, and each racetrack put in $300,000

Grandstand at Douglas Park Jockey Club, Louisville, Kentucky. *Ronald Morgan Kentucky Postcard Collection, Graphic 5, 1912, Kentucky Historical Society Digital Archives.*

into the new company. The agreement would lead to a thirty-day meet at Churchill Downs in the spring and a fall meet at Douglas Park.[25] The track was closed from 1906 to 1913. In 1913, the property was renovated, and the venue was changed to Thoroughbred racing.

Among horsemen, Douglas had friends throughout the entire United States and foreign countries. Whenever any of his horsemen friends came over to visit, he would royally entertain them. He became known as a "prince of good fellows." He also wielded political power in the local elections. Although he never sought and refused repeatedly to accept public office, he gave his time and money in many contested campaigns to aid a friend.[26]

He also helped his friends in need. According to one story, Douglas and some of his friends went on a fishing trip to Florida. One of his friends was a leading lawyer in Louisville and heavily involved in a local bank that was on the verge of collapse. One day, when the group was ready to go fishing, the lawyer held back, and Douglas asked him, "What's the trouble, old fellow?" The lawyer told Douglas that he was just informed that the bank had failed and everything he owned was wiped out, except his home, which he was prepared to sell once he got back to Louisville. Douglas asked him, "How much do you need to tide you over?" The lawyer told him he needed $35,000, which is the equivalent to $806,306 today. Douglas sat down, wrote a check for $35,000, handed it to the lawyer and said, "Now, will you go fishing?" The lawyer eventually paid Douglas back the $35,000.[27]

Douglas also contributed to institutions for the poor and the sick as well as orphans, regardless of their religious affiliations. He annually gave away thousands of dollars to worthy causes that were brought to his personal attention. He also helped in any way to aid in the development of Louisville, whether socially or in business.[28]

In May 1915, the J.J. Douglas Company was having financial problems, and in September 1915, a committee representing the creditors of the S.J. Greenbaum Company met in Louisville for the arrangements for the entire takeover of the affairs of the Greenbaum Company, the J.J. Douglas Company, the Neversink Distilling Company and the Old Pepper Spring Distilling Company. One year later, in 1916, about $70,000, which is equivalent to $1,707,408 in today's value, was ready to be distributed to the creditors of the J.J. Douglas Company, one of the subsidiaries of the S.J. Greenbaum Company, which went bankrupt. The assets of the J.J. Douglas Company were liquidated, and a sum equivalent to 90 percent of the claims was distributed to the creditors.[29]

Douglas was twice married. His first wife was Mrs. Ella Self Douglas, who died in March 1905, and afterward he had his mother and sisters move into his home. His mother died on January 7, 1916. She was ninety-one years old. He remarried after his mother's death. His second wife was Mrs. Nannie Frances Douglas. His four sisters were Ella and Tillie Douglas, Mrs. Herminia Broadhurst and Mrs. Bell Ashburn. Colonel Douglas had no children.

On January 2, 1917, Colonel James J. Douglas died in his home in Middletown from chronic intestinal nephritis. His funeral took place at his residence. He was seventy-four years old. The funeral party drove the cars from the residence to Cave Hill Cemetery, where he was buried in the family plot, Section E, Lot 30, Grave 8. The Compass Lodge of Masons was in charge of the funeral ceremonies at the grave site. His pallbearers were Dr. L.D. Mason, J.J. Kavanaugh, William Duffy, Louis and Otto Seelbach, Harry W. Brennan, Walter Kohn, J.T. Moran, Angeran Gray, William Bowe, W.T. Milton, Lawrence Cox, Charles Smith, James B. Smith and Edward H. Hayes.[30]

In February 1917, his wife, Nannie Frances Douglas, sued the Douglas estate for half of his assets and one-third of his real estate. In Douglas's will, he left her $11,000 in cash and two pieces of property. Douglas put an antenuptial contract in the will. Four sisters named in the will received $64,000.

Above and opposite: John James Douglas monument, Cave Hill Cemetery, Section E, Lot 30, Grave 8. *Photo by author.*

Douglas's wife claimed that she knew nothing of the ante-nuptial clause. She was given only $26,000 (equivalent to $542,453.44) out of an estimated $500,000 to $750,000 (equivalent to $10,174,765.63 to $15,674,695.31 in 2021 money). The sisters were defendants in the lawsuit. In August 1917, a

compromise was reached between Nannie Douglas and other heirs to the estate. Nannie would be paid a substantial sum of money in cash as long as she agreed to relinquish all claims to any and all of the real estate.[31]

By 1918, Douglas Park was no being longer used as a racetrack and became the exercise and training track for Churchill Downs; it was also used to stable Thoroughbreds that could not be accommodated at Churchill Downs. In 1939, the Douglas Park grandstand and clubhouse were demolished, and fires in the 1940s and 1950s severely damaged the horse barns. On October 26, 1952, a fire completely destroyed the largest horse barn at Douglas Park, killing sixty-eight horses. In 1954, Churchill Downs began to sell the Douglas Park property, and by 1958, the Douglas Park track no longer existed. Currently, an industrial park, homes and apartments have replaced the Douglas Park property. The only remnants left from the Douglas Park track are the brick columns leading to Holy Rosary Academy, which is located at 1327 South Fourth Street.[32] After Douglas's death in 1917, the Eothen Stock Farm (or Douglas Place) went through several owners, and his home was eventually demolished. In 1973, the farm was developed into a residential subdivision known as Douglass Hills. The only thing left from the Douglas Place farm is a springhouse.

CHAPTER 2

THOMAS SHERLEY

E.L. Miles Company and the New Hope Distillery Company

Thomas Sherley was the son of Zachary M. Sherley, a prominent steamboat captain. Thomas was born on January 1, 1843, at Fifth and Chestnut Streets in Louisville, Kentucky. In 1863, he graduated from the local schools. In May 1864, at the age of twenty-one, he married Ella Swagar, daughter of Captain Joseph Swagar, a steamboat captain. He had five children: Eva Belle (married to Robert William Lewis), Zach, Clara, Swagar and Minn-Ell. He also had a half-brother, Joseph.[33]

Sherley had many different business interests. He and his business partner E.L. Miles owned the E.L. Miles Company and the New Hope Distillery Company. He was the senior member of the firm T.H. Sherley & Son, with T.J. Bateman being his business partner. The firm of T.H. Sherley & Son was the distributor and sole agent for the two distilleries. His specialty was the sale of apple and peach brandy. He was also one of the builders of the Public Elevator Company, was one of the owners of the Kentucky Glass Company and the Southern Glass Company and had interests in the Belle of Nelson Distillery Company. He also owned land in the northwestern part of the United States, where he took his vacations.[34]

He was the former president of the Kentucky Distillers Association. He was at the forefront in fighting for the interests of the Kentucky distillers. He fought for legislation at Frankfort and Washington for the Bottle-in-Bond bill. He made frequent trips to Washington and made numerous appearances before Congress.[35] He was chairman of the Citizens Committee for the Grand Army of the Republic during the 1895 convention, which was held

AND OF NEW ALBANY, INDIANA. 257

T. H. SHERLEY ESTABLISHED 1867. T. J. BATMAN.

T. H. SHERLEY & CO.,

Commission Merchants for the Purchase and Sale of

KENTUCKY WHISKIES.

Office and Warehouse, No. 114 E. Main Street.

APPLE AND PEACH BRANDIES A SPECIALTY.

SOLE AGENTS FOR THE DISTILLERIES OF

E. L. MILES & CO. and THE NEW HOPE DISTILLERY CO.,

LOCATED AT NEW HOPE, KY.

Left: Thomas Sherley & Sons advertisement. *From* The Industries of Louisville and New Albany, Indiana *(1886).*

Right: Thomas Sherley. *Memorial History of Louisville.*

in Louisville. He worked for years to secure the Triennial Conclave of the Knights of the Templar for Louisville. The Templars recognized Thomas's suitability as a manager and made him chairman of the executive committee for the Triennial.[36]

Thomas had a deep interest in the High School Alumni Association and was vice-president of the organization for many years. He was the first president of the board of parks commissioners and served for eight years. He worked hard for Louisville streets and parks. Even after he retired, he took an interest in the parks and made suggestions to the commissioners.[37]

He was twice elected as a director of the board of trade. He succeeded Henry Watterson as Democratic National Committee member for Kentucky and served from 1892 to 1896. During the 1892 campaign, he was made a member of the subcommittee and was called to New York to confer with Democratic leaders. When he returned to Louisville, he was given an ovation by the Democrats of Louisville. When William Jennings Bryan was nominated for president at Chicago, Sherley's headquarters were at the Auditorium Annex. He made all the arrangements for the Kentucky delegation and cared for all the politicians from all parts of the state.[38]

In 1896, he was mentioned as a Democratic candidate for Congress. He did not seek the nomination. He served for six terms as a member of the Louisville School Board and was elected president. He was one of the leaders in the formation of night schools in Louisville, aided the interests of Black school students and built many schools for the Black children of Louisville.[39] He was past grand commander of the State Commandery of the Knights of the Templar and was past eminent commander of the Louisville Commandery No. 1.[40]

Thomas was also a charitable individual and gave away a great amount of money toward these endeavors. T.J. Batman, who was his partner in Sherley & Sons, opened up a charity without Sherley's knowledge. Batman keep a record of the money that went toward the charity. After Mr. Batman had kept the account for a year, Sherley looked over the books one day and ran across the charity account. He asked Batman, "What's this?" Batman answered, "That's the charity account." Sherley closed the book and set it to the side. He told Batman, "I don't want to know what's given away. We don't need the account." He never again looked at the book, but he continued to help the needy. He had many times paid off notes for people. Batman drew the notes up in a regular form, and within three or four months, the notes were to be paid in full. Sherley would pay off the notes.[41]

Thomas Sherley raised a fund for the victims of the Louisville Great Cyclone of 1893. He extended the distribution of funds to help the victims for a year, under the belief that if all the money were distributed at once, the money might be squandered, or unfeeling creditors might take the money away from those who were still in need of it. He succeeded in having his plan adopted, and the funds were distributed over a considerable amount of time, as different people needed the money. According the *Louisville Courier-Journal*, a little girl who was affected by the tornado who needed the money more than any others came to Sherley for money, but the fund had run out; Sherley did not have the heart to tell her. She continued to visit him with regularity. She never left his office empty-handed. She never knew that the money was coming from Sherley's own pocket and not from the dried-up fund.[42]

Mr. Batman told another story to the local paper about Sherley's charity. When the night schools opened in Louisville, Sherley offered a prize for the best student among the boys. He found a little boy on Main Street who was bright and capable but who had no education as well as a physical disability. He took an interest in the boy and called him into his office. He told the boy about the advantages of having an education and advised him to attend night school. He told the boy that he would pay for his schooling. The boy attended night school and won Sherley's prize, which was a silver watch. When the boy went to claim his prize at Sherley's office, he also gave the boy an encouraging talk. One day, while Sherley was walking down one of the business streets, a well-dressed, energetic young man approached Sherley and asked him if he recognized him. Sherley stated that he did not remember the young man. The young man said, "Well, I'm the young man you helped through night school. I'm prospering here and I want you to meet my family."[43] The young man took Sherley to his elegant home, introduced Sherley to his family and told them Sherley was the reason for his success. The young man was one of the owners of the largest stove manufactories in the town and one of the largest in the country.[44]

Sherley helped countless girls and boys enroll into a business college and either paid for their education or aided them in securing their education. When the young women and men left college, he always made sure that they had employment.[45]

T.J. Batman, Sherley's business partner, who started as an office boy for him in 1873, stated that Sherley never fired an employee in the twenty years he knew him. Batman stated that Sherley was too tenderhearted to discharge an employee, and when he had to let someone go, Batman was the one who gave the bad news to the employee. He did not recognize creed or color, and Catholics and Protestants alike were treated the same. Twice a year, the Little Sisters of the Poor visited Sherley's office and were given a regular allowance.[46]

In 1898, Sherley began to complain about a cold that he had for about ten days. He caught the cold when he attended the funeral of Emile Bourlier about two weeks earlier; he had also conducted the Templar service at the funeral. The day was chilly, and the service was not held until later that afternoon. He continued to attend to his business and did not send for his doctor, George Griffiths, until November 20. Griffiths told Sherley that he was suffering from indigestion and congestion of the lungs. He told him that his cold was not serious and that he was not in any danger.[47]

Sherley continued to improve and walked around his house at 207 West Breckinridge Street daily. He ran his business from his home, and on November 24, 1898, several days before his death, he dictated almost twenty letters to his stenographer. He also received numerous friends and representatives from various businesses concerns in which he was invested.[48]

On November 26, 1898, he was busy formulating the plans for the Knights Templar Triennial, with the executive committee arriving at his house and the big event set for August 27–30, 1901. On November 28, 1898, he got out of bed and asked for a bowl of cold water to wash his face. His son Swagar Sherley entered the room and found his father seated in his chair unconscious. A messenger was sent for Dr. Warner, but Sherley had already died in his chair. Mrs. Sherley, Swagar Sherley, Miss Clara Moore and Minn-Ell Sherley were called into the room.[49]

Thomas Sherley tombstone, Cave Hill Cemetery, Section P, Lot 235, Grave 5. *Photo by author.*

Thomas's regular doctor stated that he died of paralysis of the heart. His death came as a great shock to Louisville. His funeral, filled with people, was held at the Christ Church Cathedral. The chancel was nearly filled with flowers. The services were held by Bishop Dudley and Reverend J.G. Minnigerode. The coffin was almost buried under a heap of white roses and white chrysanthemums. His honorary pallbearers were J. Moss Terry, T.C. Timberlake, C.E. Dunn, John H. Leathers, Samuel Casseday, Americus Whedon, William H. Meffert and John Stratton. His pallbearers were E.L. Miles, Walter B. Haldeman, Oscar Fenley, William Cornwall, James Pirtle, Charles Gibson, Charles P. Weaver and T.J. Batman.[50]

After the funeral, the carriages left the cathedral, heading up Broadway and toward Cave Hill Cemetery. The Knights of the Templar's funeral service was conducted at the grave in full uniform. The Thomas Post and Whitaker Post of the Templars attended the funeral, along with several trade organizations.[51] Members of the Grand Army of the Republic also attended. The Kentucky Distillers Association met at the Galt House and formed a committee on resolutions to pass decrees of condolence at the loss of Sherley. Thomas Sherley was buried at Cave Hill Cemetery, Section P, Lot 235, Grave 5.

CHAPTER 3

JOHN McDOUGAL ATHERTON

Atherton Distillery

John McDougal Atherton was born in Larue County, Kentucky, on April 1, 1841. His parents were Peter Atherton and Elizabeth Mayfield Atherton. His father was a native of Virginia and moved to Kentucky when he was given a Virginia land grant of one thousand acres the year before Kentucky was admitted as a state. His maternal grandfather was Alexander McDougal Atherton, who was a soldier under General George Washington during the American Revolution. From 1800 to 1830, Peter Atherton built and operated a log distillery on the west bank of Knob Creek. When John was three years old, his father died, leaving him the estate. His mother married Marshall Key. John attended the public schools until he was ten years old, when his parents sent him to Bardstown, Kentucky, to receive an education. After attending the school in Bardstown for three years, he entered Georgetown College in Georgetown, Kentucky. In 1861, he married Maria Farnam of Georgetown, the daughter of Johnathon E. Farnam, a professor at Georgetown College. In 1862, they had a son, Peter. John did not graduate from Georgetown due to ill health.[52]

After regaining his health in 1867 and with the financial help of his stepfather, Marshall Key, John built a new distillery on the banks of Knob Creek and named the whiskey after himself. Marshall Key and John Atherton became joint owners of Atherton Distillery. In 1869, John purchased an interest in a small distillery and moved his distillery to the east bank of Knob Creek, across from his first distillery; he placed his cousin Alexander Mayfield in charge of the plant, calling his new endeavor the Mayfield Distillery. He created a village for his workers in both plants and called his village Athertonville, which was located about two miles from New Haven, Kentucky. He built his own depot as well as two miles of railroad tracks to connect the distilleries, with the main line intended to take his product to New Haven.[53] He built a large store known as "The Fair" and also built the Atherton Hotel in the 1890s in Athertonville.[54]

In 1873, at the age of nineteen, John moved to Louisville and studied law at the Louisville Law School. His health turned for the worse, and he had to move back to his parents' farm for a short time. From 1869 to 1871, he served as a member of the state legislature and was chairman of the Democratic State Central Committee. In 1881, he was elected a member of the board of directors of the National Bank of Kentucky and was later elected to serve as vice-president and eventually president.[55]

Cochran and Fulton, wholesale liquor dealers in Louisville, purchased from Marshall Key his half interest in the Atherton and A. Mayfield Distillery and partnered with John Atherton in the production and sale of whiskey. Cochran and Fulton also partnered in the Mayfield Distillery, and in 1875, Cochran, Fulton and Atherton ran both distilleries. Cochran and Fulton continued to run its wholesale liquor business in Louisville, but in 1880, Atherton became a partner in the company; the name was changed to Cochran, Fulton and Atherton. In 1880, Atherton gave Cochran and Fulton half interests in the A. Mayfield Distillery, and the three decided to build two other distilleries. By 1882, Atherton had built four distilleries, which together had the largest capacity of any company in the United States: Atherton Distillery, A. Mayfield Distillery, William Miller Distillery and the S. O'Bryan Distillery. William Miller was Atherton's bookkeeper for the J.M. Atherton & Company and the A. Mayfield & Company Distillery, S. O'Bryan was their distiller and James Maxey was a railroad agent at New Haven. Miller, O'Bryan and Maxey all owned one-sixth interest each in the company.[56] The brands made there were the Atherton, established in 1867; the Windsor, established in 1880; the Mayfield, established in 1870; and the Clifton, established in 1880. The first two were sweet mash whiskies, and the second two were sour mash brands.

Atherton's brands soon expanded to ten different labels, including Old Indian River Rye, Howard, Kenwood, Brownfield, Baker and Carter. The buildings and attached premises covered an area of thirty acres, and Atherton employed about 150 workers. When the distilleries were in full production, they consumed about 1,800 bushels of grain per day, and they produced from eighteen thousand to twenty thousand barrels of whiskey annually. He also owned his own cooperage, which employed between twenty and twenty-five skilled workers who used 600,000 staves yearly for making barrels for his whiskey. At the time, Atherton's distilleries were the only ones in the state that produced exclusively pure rye whiskey, and they were the first to make a move in the direction of rye whiskey. At the height of his distillery, Atherton was the largest producer of sour mash in the world. When he moved to Louisville, he brought the offices of his distilleries with him and sold his product to all parts of the country through the Louisville office.[57]

By 1882, Atherton had dissolved his partnerships with Cochran and Fulton, William Miller Distillery and S. O'Bryan Distillery. He also ended the partnership with Cochran, Fulton and Atherton. He paid Cochran and Fulton $80,000 for his interest in Cochran and Fulton and Atherton and his interest in the Spring Hill Distillery in Franklin County.[58]

In 1899, Atherton decided to sell his distillery business, which was at that time one of the largest and most modern plants in Kentucky, to a corporation known as the Kentucky Distilleries and Warehouse Company, or "Whiskey Trust," and he then devoted his time to real estate development and other financial interests. When he sold his business, his whiskey brand was worth $1 million, which in today's equivalent is worth $26,936,262. According to the newspaper, he sold his entire stock of seventy-five thousand barrels of whiskey. He gave up his residence at the southeast corner of Third and Broadway and sold his offices on Main Street. He built a new house for $50,000, designed by Charles Clark. Before Atherton built his house on the site, Captain Silas Miller had built a home there. Atherton finally sold his home in 1902, and in 1911, his home was demolished to make room for the Weissinger-Gaulbert building. Atherton had forty acres of land on his country estate, which was directly east of Eastern Parkway, and stated to the local newspaper that he was going to raise cattle and poultry.[59] He made real estate investments in property at the southwest corner of Fourth and Chestnut Streets, which was occupied by the Francis Building; the southeast corner of Fourth and Walnut (now Muhammad Ali) Streets; and the building erected for the Lincoln Bank and Trust Company.[60]

Atherton was the first president of the Lincoln Savings Bank and Trust Company. In 1884, he was made director of the Louisville Gas Company, and in 1898, he was elected to the board of directors for the Louisville and Nashville Railroad Company. He was also director of the Louisville Realty Company.[61]

In 1886, Atherton helped organize and served as president of the National Protective Association, which was organized in Chicago to oppose the constitutional prohibition of alcohol in the United States. He insisted on making the fight against Prohibition open and demanded that every dollar spent by the association be used to defray the legitimate expenses. The association was successful in preventing constitutional prohibition in seven successive states where the question was brought to a vote. Henry Watterson, editor of the *Louisville Courier-Journal* and a close friend of Atherton's, stated that Atherton had for the first time demonstrated the effectiveness of an open discussion type of campaign in American politics.[62]

So long as the National Protective Association was active, no state adopted the amendment prohibiting alcohol. While in the state legislature, Atherton became associated with John G. Carlisle and James B. McCreary. At the 1895 Democratic State Convention, Atherton opposed the adoption of the principle of free silver. He wrote and spoke against the free silver plan and was a delegate to the Democratic Sound Money Convention in Indianapolis, Indiana.[63]

Atherton opposed the 1891 Kentucky Constitution on the grounds that the constitution's provisions of taxation were unequally distributed between real estate and personal property.[64]

Atherton also had an interest in education. In 1884, he became a member of the Louisville Board of Education. He was instrumental in having the old school trustee law abolished for Louisville and the Louisville school system put under a board of education. In 1893, he made a $30,000 donation to Georgetown College and stated in his letter to the president of the college that the money was to be used as an endowment, to be named the Atherton-Franam Chair of Natural Science. He stated that he was educated at the college and that his wife's father spent nearly fifty years as a professor at Georgetown in the Natural Science Department.[65] In 1910, he also served as chairman of the board of trade's committee, which selected candidates for the city's first nonpartisan board of education. On December 10, 1921, after suspending a rule forbidding the naming of a school after a living person, the Louisville Board of Education decided to name the proposed new girl's high school on Morton Avenue at Rubel after Atherton.[66]

Atherton was also one of the founders of the Pendennis Club. He was also a former member of the Louisville Jockey Club, which would become Churchill Downs. He was also a former member of the Louisville Country Club.[67]

On May 30, 1932, Atherton came down with pneumonia and died several days later on Sunday night, June 5, 1932, at his home on 2542 Ransdell Avenue, Louisville. His funeral was held at his home, and services were conducted by Reverend Dean Richard L. McReady of Christ Church Cathedral. His active pallbearers were Clifton Atherton, John Atherton Miller, Edmund Miller, Churchill Humphrey, Owsley Brown and Thomas Floyd Smith.[68] He was buried at Cave Hill Cemetery, Section 13, Lot 110, Grave 2.

CHAPTER 4

SAMUEL GRABFELDER

Echo Springs and Rose Valley

Samuel Grabfelder was born in Rehweiler, Bavaria, Germany, on September 2, 1846. His parents were Emmanuel and Regina Grabfelder. He was educated in the local schools in Germany. In 1856, when he was still a boy, he came to America with his family, and in 1857, they settled in Louisville, Kentucky. He attended Louisville Male High School, where he learned the English language and business skills. With these skills, his industriousness and his general intelligence, he had no problems securing employment. By the close of the Civil War, he had developed into a capable salesman and businessman. For several years after the Civil War, he was a traveling salesman working for one of the large wholesale liquor houses in Louisville. He traveled mostly across the southern states. On February 13, 1870, he married Cordelia Griff of Louisville. While employed with the company, he saved his money, and in 1873, he became a partner at L. Oppenheimer & Company of Louisville, which was a wholesale liquor company.[69]

In 1879, after six years, he cut his connections with Oppenheimer and opened his own liquor wholesale business called S. Grabfelder & Company Whiskies. He relied on several distilleries to supply his rectifying operations, which blended different bourbons, including Pleasure Ridge Park Distillery, the Mayfield Distillery and the Crystal Springs Distillery—all of which were located in Louisville.[70]

In 1893, Grabfelder was offered a contract to handle all the sales from the Murphy-Barber Distillery. The distillery was owned by Squire Murphy, M.

Barber and Calvin Brown. The distillery had been in operation for about thirteen years. During that time, Murphy died and Brown decided to retire, offering Grabfelder a contract to handle all of the sales of the distillery. The distillery was located in Bullitt County about twenty miles from Louisville on the Bardstown branch of the Louisville and Nashville Railroad. The distillery made two hundred bushels of grain per day and was one of the best small distilleries in Kentucky. In 1899, Samuel Grabfelder bought the Murphy-Barber Distillery for about $30,000. The warehouses held twenty-five thousand barrels. All the brands were transferred to Grabfelder.[71] Samuel, along with his brothers Morris and Mose Grabfelder, located his main office on Main Street, between First and Second Streets in Louisville. Now that he owned a distillery, Grabfelder sold a number of brands, including Cane Spring, Clermont Rye, Dan Becker's Monogram, Dunn's Monogram, Horse Shoe, Kentucky Belle, S. Grabfelder's American Malt, Southern Pride and Woodford County. His flagship brand was Echo Springs and Rose Valley.

Grabfelder's bourbon business made $50,000 the first year but gradually increased to making millions every year. Traveling men represented his liquor business in every state, and his famous brands of whiskies were known throughout the country. Among the various liquor houses engaged in the distribution and sale of Kentucky whiskies, none had a higher standing than S. Grabfelder & Company Whiskies. In 1905, Grabfelder sold his home, which was located on 1442 Third Street, to W.H. Kaye Sr. He and his wife were moving to Philadelphia, which was where his wife's family was located, and later he settled in Atlantic City, New Jersey. His home cost $100,000 to build and was located on one of the most fashionable blocks of the residential district on Third Street in Louisville.[72] The home was made of white stone and was one of the most perfect residences ever built in Louisville during that time.

In 1909, S. Grabfelder & Company Whiskies moved to 119 West Main Street. He ran the company from Philadelphia until 1919, around when federal Prohibition was enacted. During Prohibition, Grabfelder liquidated his business, and his distillery was torn down, except for one warehouse. The old distillery grounds eventually became the site for the present-day Jim Beam distillery. Although he lived in Atlantic City, Grabfelder still remained a member of the congregation Temple Adath Israel in Louisville.[73]

Samuel Grabfelder was a member of the board of trade, the Commercial Club and the Standard Club. He was also a Mason of the 32nd Degree. He was Jewish and was president for several years of the Temple congregation of Louisville. For several years, he was president of the Old Folks home of

Cleveland, Ohio, which was a Jewish charitable organization. He was the president and founder of the National Jewish Hospital for Consumptives, which was built in Denver, Colorado, at a cost of $250,000. In 1905, he founded the Jewish Free Hospital in Louisville. He was also the owner of real estate in Louisville occupied by the Raffo Furniture Company.[74]

On April 17, 1920, Samuel Grabfelder died at his home in Atlantic City, New Jersey, and was buried at Salem Fields, Brooklyn, New York. He was survived by his wife, Cordelia Griff Grabfelder, brother Morris and three nephews: Moses, Robert and Abe. He was estimated to have been worth between $5 million and $6 million. In his will, he gave $50,000 to the National Jewish Hospital for Consumptives in Colorado and $10,000 to the Jewish Orphan Asylum in Cleveland. He gave $100,000 to his wife, Delia, and created a trust fund of $500,000. He gave $10,000 to the Jewish Hospital Association of Louisville and $10,000 to the Jewish Charities of Louisville. He also gave $2,000 to the Jewish Sheltering Home in Philadelphia. His residentiary estate was valued at $1,125,000 and was to be divided among his relatives.[75]

His nephew Emmanuel Grabfelder, who was his treasurer, died on October 28, 1906. His brother, Morris, died on August 7, 1920, at his home in New Jersey. Moses Grabfelder, who was the distiller for Samuel Grabfelder and president of the Clermont Cafeteria Corporation, died at his home on 1521 South First Street on May 18, 1924. He was buried at Adath Israel Cemetery. He was born on November 5, 1865, and was connected with S. Grabfelder & Company until Prohibition. After the liquidation of the company, he became president of Jewish Hospital, the Federal Jewish Charities and the Standard Club. He was a Mason, Shriner and Elk member and a member of the Order of B'nai B'rith. Toward the end of his life, Moses was in the cafeteria business at 423 West Market Street.[76] Abe Grabfelder, who was Samuel Grabfelder's accountant, died on September 24, 1929, at his residence at the Brown Hotel.

CHAPTER 5

DAVID H. TAYLOR AND J.T. WILLIAMS

Yellowstone, Rich Hill and Honey Dew

David Hart Taylor was born on April 1, 1840, in Henderson, Kentucky. When he was a child, his family moved to Louisville, and he was soon in business with his relatives until the Civil War. On August 15, 1862, he enlisted in the ranks of the Confederacy. From September to October 1863, he was present on the muster rolls as a private in Company C, Second Battalion Cavalry (Confederate States), under Captain John B. Dortch. The Second Consolidated Cavalry Battalion was assembled in August 1863 and contained men who had either escaped capture or did not go on Confederate general John Hunt Morgan's 1863 Indiana-Ohio Raid. Morgan was captured along with most of his command at New Lisbon, Ohio. Colonel Adam Rankin Johnson reorganized Morgan's men into two separate battalions. The men from the First Brigade were organized into Kirkpatrick's First Battalion Kentucky Cavalry, under the command of Captain John D. Kirkpatrick. The men from the Second Brigade were organized into Dortch's Second Battalion Kentucky Cavalry, under Captain John B. Dortch, formerly of the Seventh Kentucky Cavalry. The Second Battalion was assigned to General Basil Duke's Brigade in the Department of Western Virginia and East Tennessee and saw action in several conflicts in Kentucky, Tennessee and Virginia. By January 1865, the battalion had only thirty-two men left and disbanded. During the Civil War, Taylor was captured and was a prisoner of war in Northern Ohio.[77]

When the Civil War ended, Taylor worked in Clarksville, Tennessee, for about a year and then moved back to Louisville. He was in business with E.H. Chase & Company, wholesale liquor dealer. In 1875, he started a business of his own and, in 1878, formed a partnership with John T. Williams under the name Taylor & Williams, which was located at 107 West Main Street.[78] John T. Williams was born on August 10, 1849, in Dixon County, Tennessee. He received his education from his local schools and moved to Nashville, Tennessee, where he was a bookkeeper for a local grocery. He saved his money and soon had enough to open a modest business of his own. He developed his own fortune, moved to Louisville in 1872 and became full partner with Taylor. The two men owned the Nelson County Distillery, which made Yellowstone, Rich Hill and Honey Dew Whiskey. The leading brand, Yellowstone, was the idea of Charles W. Townsend, who was one of the partners in Taylor & Williams. Townsend was present at the dedication services at Yellowstone National Park, and he thought Yellowstone would become an outstanding American name for a whiskey. When he returned to Louisville, he insisted that Taylor & Williams register the brand Yellowstone. The brand immediately became popular, particularly in the West.[79]

On July 1, 1892, the partnership ended when Taylor, who had been financially stressed in his business enterprises, retired from business completely, and Williams became president of the company. Interestingly, Williams never drank his own whiskey. He also advised other men not to drink his product. He was a heavy smoker though.[80]

In 1870, Williams married Laura Butler of Nashville. They never had any children. He often stated that much of his success in life was due to his wife. Williams had three brothers, all of whom lived in Tennessee. He was a supporter of the free silver policy and was one of the few Main Street merchants who supported William Jennings Bryan's candidacy for president in 1896.[81]

In 1894, David Hart Taylor developed a severe cold that turned into pneumonia, and this eventually led to his death on March 16, 1894. He died at his residence on 1230 Fourth Street. His funeral was held at his home; he was a member of the Elks, and some of these men acted as pallbearers.[82] He was laid to rest at Cave Hill Cemetery, Section E, Lot 177, Grave 22. On February 22, 1903, John T. Williams died at his home on 110 West Ormsby Avenue. His death was caused by a nervous collapse induced by asthma and bronchitis. He had amassed a large fortune. He was laid to rest at Cave Hill Cemetery in a temporary vault until Mrs. Williams was able to accompany her husband's remains to Nashville, where he was laid to rest at Mount Olivet Cemetery, where she had a family cemetery plot. Some of his pallbearers were distillers: Charles P. Moorman, Joseph Bernard Dant and

David Hart Taylor monument, Cave Hill Cemetery. *Photo by author.*

John Weller. Other pallbearers included Thomas Satterwhite Jr., William M. Morris, Thomas Timberlake, Thomas Batman and C.H. King.[83]

Williams made his will on his deathbed, but his two brothers contested the will. He bequeathed to his one brother at Union City, Tennessee, a lifelong interest in a hotel. To the other two brothers, he left a tract of unimproved land in Dickson County, Tennessee. The remainder of his property was to

J.W. Dant Bourbon tip bottle. *Ryan Reynolds Collection.*

be divided among his heirs and in-laws, with the exclusion of the two brothers—under the Commonwealth of Kentucky laws, his widow would receive half of the personal property, and three half brothers, a half sister and a nephew also had shares in the distribution of his wealth.[84]

The property was reduced to money except for the unimproved land. The cash bequests amounted to $250,000, and the Tennessee property would bring the total to $300,000. A judge arbitrated a compromise: one brother was given ownership in fee of the hotel, and concessions would be made for the other brother, to whom the Dickson County property was left.[85]

When John T. Williams died, the business was to be conducted by the executor of the estate, the Fidelity Trust and Safety Vault Company. Just before his death, he gave two employees of the company interest in the business. C.W. Townsend, a salesman for the company, and W.H. Duane, manager of the company, also had an interest in the company. The Fidelity Trust and Safety Vault Company would have to incorporate the company. Williams also had a controlling interest in the Charles P. Moorman Company. Moorman planned to buy back his stock.[86] At the turn of the century, Joseph Bernard Dant became president of the Taylor & Williams Distillery.

During Prohibition, Yellowstone was bottled and sold as medicinal whiskey. For seventy years, J.B. Dant made Yellowstone. Dant and his wife, Nannie, owned the entire capital and stock of Taylor & Williams. In 1939, Dant died, and he was buried at Calvary Cemetery in Louisville. In 1943, the Fidelity Trust and Columbia Trust Company was entrusted to sell Taylor & Williams Inc. and Taylor & Williams Distillery. The distillery companies had forty thousand barrels of whiskey maturing in warehouses. The barrels were worth $100 per barrel. With forty thousand barrels, the value was $4 million.[87] By 1944, the Dant Distillery, along with the Taylor & Williams distillery, was making six hundred bushels daily and sixty barrels of bourbon per day. The still, the six-thousand barrel warehouse and a bottling house with the capacity of making 2,400 cases daily were located at 7000 Seventh Street Road. In 1944, the president of Taylor & Williams distillery was Michael J. Dant, and other officers included Samuel Dant, J. Walter Dant and James Kearns. Glenmore Distillery, which began under James Thompson, bought all of the stock of Taylor & Williams, as well as the remaining stock of the Taylor & Williams Distillery. The holding company was part of the estate owned by J.B. Dant and his wife, Nannie Dant. Glencoe bought Taylor & Williams and other assets of the Dant family for between $5 million and $6 million.[88]

CHAPTER 6

CHARLES MOORMAN

J.H. Cutter Old Bourbon

Charles P. Moorman was born in Breckinridge County in 1829 and was educated in the local schools. In 1845, he moved to Louisville and became a bookkeeper for the J.H. Cutter Company. By 1859, he was a business partner with Cutter. Cutter and Moorman had their business office in Louisville and the distillery in Bourbon County. Moorman made his first fortune by sending a shipload of Cutter whiskey around Cape Horn to San Francisco, California. At the time, the whiskey started at a tax of ninety cents per gallon, which had been enacted by the federal government. By the time the cargo reached San Francisco, the tax had been raised to two dollars per gallon, giving him a wide margin on the shipment.[89] He sold his Cutter whiskey to A.P. Hotaling Company in San Francisco and made a profit of fifty dollars per barrel. His offices for J.H. Cutter Old Bourbon Whiskey were located at 131 and 133 East Main Street. He also had branch offices where sales were made direct in Boston, New York, Philadelphia and San Francisco.

During the Civil War, Moorman spent a large part of his time in California and the rest in Europe. In 1865, he married Lucy Beckley of Shelby County. He had two sons: C.P. Moorman Jr. and Elmore. Elmore died in 1902. By 1872, John Cutter had passed away, and Moorman was now sole owner of the distillery.[90]

Although he was worth at least $2.5 million, he lived in simplicity in his home with his son, C.P. Moorman Jr., and his housekeeper, Mathilda Jennings, who had been a servant in the Moorman family for thirty-eight years. He spent time with his nephew, Arthur Board, at the Louisville Hotel, and they both would take business trips downtown.[91]

Moorman was also president of the Cloverport Brick and Manufacturing Company. In 1894, he retired from the whiskey business and sold his holdings to W.C. Wheeler and Alex Semple, who were business associates. Semple died, and Wheeler bought the remaining stock in the business.[92] While head of the Cutter distillery company, Moorman elevated the Cutter brand of whiskey to one of the most famous in the country. In 1912, Wheeler, who was president of the C.P. Moorman & Company, bought the W.B. Samuels distillery—which was located at Samuels Depot, Nelson County, on the Louisville and Nashville Railroad—and expanded the capacity of Cutter bourbon.[93]

After retiring from the whiskey business, Moorman invested his money in securities and became one of the city's best-known capitalists. He had interests in various businesses in Louisville, including woolen mills, and was one of the largest local stockholders in the Louisville Railway Company.[94]

In 1912, Moorman decided to retire from the Fidelity Trust Company. He had served as a member of the company for twenty-seven years. He held 1,200 shares in the company and decided to sell them because he stated that he was not taken into confidence by some of the directors prior to the announcement of the proposed merger between the Fidelity Trust Company and the Columbia Trust Company. All of his stock was bought by a syndicate of local bankers and capitalists who were hoping to consolidate the two trust companies. He sold his stock for $190 per share, for a total of $237,000.[95]

Although he never talked to any of his family about his business and never referred to his own affairs, Moorman was one of the most lavish in Louisville when it came to charity. No one was ever turned away from his door. He always came to the assistance of the Little Sisters of the Poor.[96]

On February 13, 1917, he attended to his various business affairs downtown, including signing his income tax statement, and returned to his home; by 3:00 p.m., he had suffered a massive heart attack at his home on Fourth Street. He died at the age of eighty-six and was laid to rest beside his wife, who had died in 1891, at Cave Hill Cemetery,[97] Section A, Lot 190, Grave 3. His wife had died while on a cruise on the Cunard line steamship *Umbria*.

In his will, Moorman made the largest bequest ever made by a resident of Louisville to charitable purposes. He left half his fortune to founding a home for the indigent women of Louisville and Jefferson County—or any other old woman who might come into the county and be a resident for a period of five years before asking admittance to the home. The home would be called the Charles P. Moorman Home for Women. The rest of the estate went to his invalid son, Charles Moorman Jr., and his granddaughter, Lucy Elizabeth Moorman. His housekeeper, Mathilda, was given $10,000. Joan Moorman, widow of Elmer Moorman, was given $50,000. Arthur Board was given $20,000. His nieces Ella Moorman, Martha Smith and Emma Haswell were given $2,000 each, Sophronia Moorman was given $1,000 and the Little Sisters of the Poor and Jewish Hospital were given $5,000 each.[98]

Lucy Elizabeth Moorman contested the will, and in 1920, the courts made the decision that the millions that were supposed to go the aged women home would not become a reality. Lucy Elizabeth Moorman Bartlett had a daughter, Joan Elizabeth Bartlett, who would become the heiress of C.P. Moorman's millions.[99] But not all hope was lost for the indigent home. In Moorman's will, he left half his estate to his invalid son, Charles Jr., and upon his death, half of his estate would be used to purchase property in Louisville to provide a home for the poor, the indigent and aged women. Moorman Sr. designated Leon Solomon, Caldwell Norton and Arthur Board as a committee to manage the estate for his son after his death and to establish the home. Charles P. Moorman Jr. died in 1925, leaving an estate worth more than $1 million, and the committee agreed to open a home on 245 Chestnut Road and Highland Avenue and later build a new home on 966 Cherokee Road. The home had a capacity for sixty women.[100]

Just recently, the A.P. Hotaling Company, formerly the Anchor Distilling Company, brought back the J.H. Cutter Bourbon brand. According to the company, the Cutter brand is made up of a blend of whiskies, with 73 percent Kentucky bourbon aged three to five years, 17 percent Old Potrero 18th Century Style Rye Whiskey aged two and a half years and 10 percent Old Potrero Port Finish Rye Whiskey aged two and a half years to three years. The bourbon sells today for fifty dollars per bottle.[101]

CHAPTER 7

JOHN G. ROACH

Old Log Cabin, Rich Grain, Bel Air and King of Kentucky Distilleries

John G. Roach was born on March 29, 1845, in Campbellsville, Taylor County, Kentucky. He was the son of John J. and Marth P. White Roach. His father was originally from Charlotte County, Virginia, and his mother was from Green County, Kentucky, the daughter of William P. White, who was a representative for Green County in the Kentucky General Assembly in 1823–24. He attended the local schools and entered Georgetown College. While he was a student at Georgetown, he converted to the Baptist faith.[102]

When the Civil War broke out, Roach left college and joined Confederate general John Hunt Morgan's cavalry. He was a member of the Third Kentucky Cavalry, under the command of Colonel Richard Gano. He participated in a number of Morgan's raids and spent most of his time on special duty. In July 1863, he was part of a detachment of Confederate soldiers under Captain Thomas Henry Hines that captured the transports *Alice Dean* and *McComb* at Brandenburg, Kentucky, which carried Morgan and his men across the Ohio River into Indiana. On July 19, 1863, he was captured at Buffington Island, West Virginia. Immediately after his capture, he was imprisoned at Columbus, Ohio, and six weeks later, he was transferred to Camp Douglas, the Federal prisoner of war camp in Chicago, Illinois. He remained a prisoner of war until 1865, when he was paroled and sent home in Kentucky. His family suffered heavy losses. He had seven cousins who had entered the Civil War, but only two survived the war.[103]

After the Civil War, Roach returned to Kentucky, found a job in the coal mining operations in Pulaski County and was engaged in mining and shipping coal from Pulaski County to Nashville, Tennessee, until 1869, when he moved to Louisville and started a wholesale whiskey trade as a member of the firm Grove, Roach and Company. The company ran until 1874, when he bought out the interest of his partner, Grove, and changed the name to John G. Roach and Company. In 1879, he sold his merchant business and turned his attention to building and operating distilleries. He built four large distilleries: one at Uniontown, Kentucky; one in Portland; one at the corner of Twenty-Eighth and Broadway; and one at the corner of Thirtieth Street and Garland Avenue. The Uniontown Distillery had the capacity of six hundred bushels, and his Portland distillery had the capacity of two hundred bushels. He sold the Uniontown Distillery to a Philadelphia syndicate and later sold his distillery at Portland and the one at Twenty-Eighth and Broadway. The distillery at Thirtieth and Garland was built in 1892 and was the largest and best-equipped distillery in Kentucky. At the distillery, he made the famous brands Old Log Cabin, Rich Grain, Bel Air and King of Kentucky Distilleries. He produced one hundred barrels of whiskey per day.[104]

Bel Air Distillery. *From* The Industries of Louisville and New Albany, Indiana *(1886)*.

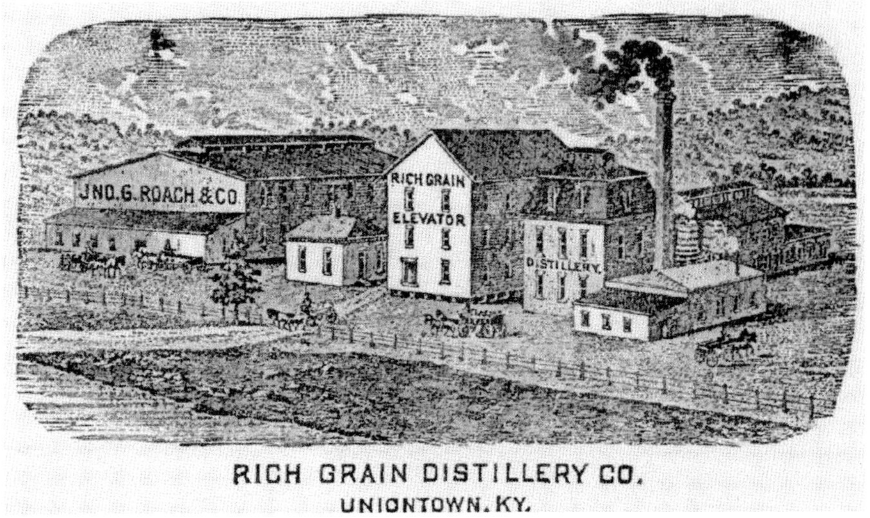

Rich Grain Distillery, owned by John Roach. *From* The Industries of Louisville and New Albany, Indiana *(1886).*

In 1899, he sold his large distillery at Thirtieth and Garland to the Kentucky Distilleries and Warehouse Company for $500,000, which amounts to $14,458,253 today. Along with the distillery were the copper shop, receiving room, warehouse and grain elevator. The plant had the capacity of one thousand bushels per day.[105] In 1906, fire completely destroyed the John G. Roach bottling plant, along with several hundred dollars of stock that was in the process of being bottled. The loss was estimated at $5,000.[106]

Roach was also prominently identified with banking and commerce as a stockholder and director in the Union National Bank, the Bank of Commerce and the Louisville Insurance Company. He was also involved in politics. From 1880 to 1892, he was chairman of the Democratic City Committee of Louisville and conducted several campaigns for his party. He was a delegate to the National Democratic Convention of 1884 and supported Grover Cleveland. He was also commissioner for the Central Insane Asylum.[107]

In 1870, he married Sally Neill, daughter of Mathew and Eliza Neill of Louisville, and they had two sons: Neill and Ethric. Neill graduated from Louisville High School, and after spending some time abroad, he returned home to become a manager of the large distillery that his father owned. John Roach died on December 9, 1907, at his apartment, The Parsons,

John Roach monument, Cave Hill Cemetery. *Photo by author.*

John G. Roach marker, Cave Hill Cemetery. *Photo by author.*

on Bonnycastle Avenue in Louisville. The funeral services were held at his residence. Services were conducted by Reverend George Eager of the Southern Baptist Theological Seminary.[108] On December 12, 1907, he was buried at Cave Hill Cemetery, Section 14, Lot 276, Grave 1. He had a private funeral. At the time of his death, he was worth $250,000. He left his estate to his wife, Sally Roach.[109]

CHAPTER 8

PAUL JONES JR.

Four Roses

Paul Jones Jr. was born in Lynchburg, Virginia, on September 6, 1840. His father was Paul Jones, and his mother was Lavinia Cary Pankey Jones. Warner Jones, brother of Paul Jones Jr., was born on October 20, 1829, in Campbell County, Virginia, and before the war, he was an attorney for the County of Obion, Tennessee. During the Civil War, Paul Jones Jr. and Warner Paul enlisted in the Confederate army and joined the Thirty-Third Tennessee Infantry. The regiment was organized on October 18, 1861, at Union City, Tennessee, for three-year service. Lysander Adams stated that he was "appointed" colonel by General Leonidas Polk in the summer of 1861 to raise a regiment and that he was instrumental in raising this regiment, but he was not elected colonel of the regiment at the election of officers. Colonel Alex W. Campbell became colonel of the regiment, and he stated that the regiment remained in Camp of Instruction near Union City until January 1862, when the regiment moved to Columbus, Kentucky. Only a few of the companies were partially armed, mostly with shotguns and hunting rifles, and the regiment was not completely armed until a few weeks before the Battle of Shiloh, when the men obtained some flint and steel muskets as a loan.

At the Battle of Shiloh, April 6–7, 1862, the regiment was in Brigadier General Charles Clark's Division, Brigadier General Alexander P. Stewart's Brigade, composed of the Thirteenth Arkansas; the Fourth, Fifth and Thirty-Third Tennessee Infantry Regiments; and Stanford's Battery. In the Battle of Shiloh, the Thirty-Third and Fifth Tennessee were commended by both

General Polk and General Stewart for their charge that led to the capture of Federal general Benjamin Prentiss. During this charge, Colonel Campbell was severely wounded but retained command. The regiment reported 20 killed, 103 wounded and 17 missing. On May 8, 1862, the regiment was organized. Warner P. Jones became a colonel of the Thirty-Third Tennessee Infantry and fought in the Battle of Perryville, Kentucky, on October 8, 1862. The Thirty-Third Tennessee fought at the Battle of Chickamauga, and during the Siege of Atlanta, on June 30, 1864, Warner Jones was killed and was buried at the Oakland Cemetery in Atlanta. Paul Jones Jr. also served in the Thirty-Third Tennessee Infantry as a second lieutenant. In September 1862, he was serving as adjutant of the regiment and later was assigned to field and staff duty. In March 1864, he was absent on post duty. In August 1864, he was ordered by Confederate general Braxton Bragg to be in the field near Atlanta, Georgia. On May 1, 1865, the Thirty-Third Tennessee surrendered and was paroled at Greensboro, North Carolina.

After the Civil War, Paul Jones Sr., along with his son Paul Jones Jr., moved to Atlanta, where they both became successful in the whiskey business. At the age of seventy-six, Paul Jones Sr. died on April 3, 1877, and was buried next to his son Warner at the Oakland Cemetery in Atlanta. In 1883, a strong temperance movement brought prohibition to Georgia, so Paul Jones Jr. moved to Tennessee, where he continued to develop the whiskey business as a broker for various distillers. In 1884, Paul Jones decided to move to Louisville, Kentucky. In 1888, Jones bought the assets from whiskey maker Rufus Mathewson Rose, who named Four Roses after himself, his brother Origen and his two sons.[110] According to the Four Roses website, run by Kirin Brewery of Japan, the legend of the name began

> *when Paul Jones, Jr., the founder of Four Roses Bourbon, became smitten by the beauty of a Southern belle. It is said that he sent a proposal to her, and she replied that if her answer were "Yes," she would wear a corsage of roses on her gown to the upcoming grand ball. Paul Jones waited for her answer excitedly on that night of the grand ball.…When she arrived in her beautiful gown, she wore a corsage of four red roses. He later named his Bourbon "Four Roses" as a symbol of his devout passion for the lovely belle, a passion he thereafter transferred to making his beloved Four Roses Bourbon.*[111]

Jones acquired the Four Roses name, along with other Roses assets, and operated the company under the name of Paul Jones Company, which was

Paul Jones office in Louisville, Kentucky. *From* Wine and Spirit Bulletin *(1903).*

located on 136 and 138 Main Street on Whiskey Row.[112] The building was four stories tall and had a basement. He handled all the leading brands of fine Kentucky bourbon and sour mash whiskies and carried large stocks in store and in bonded warehouses. In addition to handling fine products from Kentucky, he also handled the superior Pennsylvania rye whiskies manufactured by Hannes, Moore and Sinnot, successors to John Gibson, Son & Company, which were recognized as the leading brands of Pennsylvania rye. Jones was recognized as one of the most prominent men identified with the distilling and wholesale whiskey trade.[113]

Paul Jones advertisement. *From* Wine and Spirit Bulletin *(1903)*.

Jones became president of the J.G. Mattingly Company at Thirty-Ninth and High Streets. He was director and vice-president of the American National Bank. He was also a member of the board of trade and director of the National Building and Loan Association.[114]

In 1892, Paul Jones registered the Four Roses trademark, and his name and brand soon became known throughout the country. He believed in advertising, and in New York he rented space on a building at Madison Square for a sign of incandescent electric lights at a cost of $1,200.[115]

Paul Jones Jr. never married and made his residence for a number of years at the Galt House. He died on February 23, 1895, in Louisville from "abscesses of the brain." He was surrounded at his deathbed by his three nephews: Lawrence, Sanders and Bland Ballard Jones. At the time he died, he was one of the wealthiest and most widely known distillers in Kentucky.[116]

His funeral was held at the Calvary Church. The flower arrangements were profuse, with many in handsome designs—among them were a large shaft of wheat, signifying that he lived to be of seventy years of age, and a broken wagon wheel, which symbolized the circle of life, broken by death. The remains were taken to Cave Hill Cemetery, where they were temporarily placed in a public vault until a mausoleum could be built by his family. The casket was one of the most expensive and handsome seen in the city at that time, made of solid red cedar covered with imported broadcloth. The handles were oxidized silver, and the knobs were made of solid gold. His mausoleum vault was done in a Greek Revival style, and the interior contains a beautiful stained-glass window.[117]

His nephews Lawrence and Sanders Jones continued to run the family business. Jones relatives bought thousands of barrels of bourbon and brokered them. Prohibition changed the focus of the company, after which it began to make medicinal whiskey. After Prohibition, Paul Jones Company

Paul Jones advertisement. *From* Wine and Spirit Bulletin *(1903)*.

Paul Jones Jr. mausoleum, Cave Hill Cemetery, Section 5, Lot 1-Vault, Grave 7. *Photo by author*.

The beautiful stained-glass window in Paul Jones's mausoleum. *Photo by author.*

bought the Frankfort Distilleries, where the company bottled Paul Jones and Old Prentice until World War II. In 1943, Canadian distiller Joseph Seagram & Sons Inc. purchased the Frankfort Distilleries, Paul Jones and Company and other distilleries, including Old Hunter, Lewis Distillery, Athertonville Distillery, Henry McKenna Distillery and Old Prentice Distillery. Seagram moved the Four Roses brand to Europe and Asia. In 2002, the Kirin Brewery Company bought the Four Roses bourbon brand trademark and production facilities. After the new company was named Four Roses Distillery LLC, Four Roses was once again sold in the United States.[118]

CHAPTER 9

PHILIP HOLLENBACH

Glencoe Distillery

Phil Hollenbach was born in Gau-Algesheim, not far from Bingen, on December 4, 1851. In February 1869, at the age of eighteen, he moved to America and arrived in Louisville in 1873. After working for four years in a subordinate position, he opened his own place. In December 1877, Phil, along with his brother, August, founded Hollenbach Brothers. His new company was located on Market Street between Fifth and Sixth Streets on the north side of the street. The new venture was a wine business. After a year, the brothers realized that Louisville was not ready for a wholesale wine business and that the business needed a retail branch.[119]

The brothers decided to open a wine room, or *weinstube*, in connection with the wholesale business and moved their business to First and Market Streets into the cellars of the German National Bank. They established a place called *Der Bremer Rathskeller*. In February 1881, August left the business in order to make room for his brother, Louis. The business continued to grow, and eventually the business was ranked as one of Louisville's best wine houses. The business expanded outside Louisville and added whiskey, which became the main liquor carried in the cellars.[120]

In the spring of 1881, Philip made his first trip to Frankfort, Germany, and was very successful in taking several orders. In the spring of 1882, Phil traveled to Chicago, Illinois, with great success, and then he visited Danville, Decatur and other cities in the state. He also visited Indiana, including Lafayette, Bedford, Mitchell and New Albany. He visited Owensboro, Kentucky, and Milwaukee, Wisconsin. The business continued to grow, and Phil soon

Glencoe Distillery advertisement. *From* The Industries of Louisville and New Albany, Indiana *(1886)*.

opened agencies in Denver, Colorado; Los Angeles, California; and Belleville, Texas—"Glencoe Whiskey" became a staple article and famous brand in all those states, as well as in the West.[121] Glencoe Whiskey was made in the distilleries of Stitzel Brothers, whose agency Hollenbach Brothers had acquired. Hollenbach Brothers began to import wines directly from Europe and dealt with wine houses in California and Ohio.[122]

On April 1, 1884, the business outgrew its location and moved to a larger location on Second Street, between Market and Main. Louis Hollenbach left the firm in 1885 and moved to Chicago to form a business partnership with Charles Stubenrauch; he purchased the wholesale and retail wine and liquor business of Val. Haas, 128 LaSaile Street, near Chicago and changed the name to Louis Hollenbach & Company. Nace Vetter became Phil Hollenbach's business partner. Hollenbach changed the name to Hollenbach & Vetter, but he left the business two years later in order to return to politics. Vetter was a city marshal.[123]

On February 1, 1889, the business moved to the corner of Sixth and Main Streets but soon moved again to Main Street, between Sixth and Market Streets. The business continued to grow, and Hollenbach had to buy the adjoining four-story building. A few years later, he bought another one, and the corner building was abandoned. The three huge cellars and the two large buildings were headquarters for the firm for twenty-two years. The business finally settled on 528 West Main Street. The building was five stories high and contained one of the finest cellars in the entire South. The cellar measured 197 feet across and 28 feet in width.[124]

On July 1, 1902, the firm Hollenbach & Company was incorporated under the laws of Kentucky with a capital of $75,000, and the name was changed to The Phil Hollenbach and Company Inc. The incorporators were Phil Hollenbach Sr., Louis Hollenbach and Edward Oesterritter, who joined the firm in 1892 at the age of fifteen. Oesterritter became vice-president of the company. He was the son of a retail liquor dealer. Louis Hollenbach studied the wine business in Germany and lived in Germany for two years in Frankfort on Main and Bingen on Rhine, where he learned in the cellars of the business M. Heymann Sons. Under the guidance of his father, he became an expert in wines and was the secretary and treasurer of the

Matchbook advertisement for the Fortuna brand. *Author's collection.*

business. When the company was incorporated, his son, Phil Hollenbach Jr., and Charles McDonald became partners. Phil Jr. left the business to return to New York, where he joined a financial institution.[125]

The Phil Hollenbach & Company Inc. also owned one of the finest distilleries in the state. In the summer of 1902, Stitzel Brothers Company went up for sale. Hollenbach organized some men in the same line of business in St. Louis, Missouri, Chicago and Elgin, Illinois, and started the Glencoe Distillery, which was located on Twenty-Sixth Street and Broadway.[126]

Jacob Stitzel was raised in the distillery business, and his business became a huge success. He retired in 1910, and his son, Frank Stitzel, became the manager of the distillery. Phil Hollenbach was alternately president, secretary and treasurer of the Glencoe Distillery, and all the big purchases (barrels, grain and so on) were handled by Phil Hollenbach and Frank Stitzel. Frank Stitzel was educated by his father, and he became known as one of the best-known experts in the art of distilling whiskey. He started at the bottom of business and took a course in a distilling class in Milwaukee, learning how to be a chemist. Two new warehouses were built, and along with the other three warehouses, the Glencoe Distillery could store between fifteen and twenty thousand barrels. New machinery was installed. Only the best grain and

This page: Glencoe Distillery matchbook advertisement. *Author's collection.*

malt were used in the production of the whiskey, and the new product was known by the brand name Fortuna. They also made Glencoe Sour Mash, Stone Hill Bourbon and Hollenbach Rye. In 1911, the business sold more than twenty thousand cases, and the bottling plant had to be enlarged.[127] Hollenbach also made the famous Red Rose Claret and introduced Double Comet Gin to the market. Double Comet Gin was made by a Dutch distiller in a distillery in Louisville. The company also imported wines and every type

Phil Hollenbach family monument. *Photo by author.*

of liquor from around the world and was a representative for the Anheuser-Busch Brewing Company Association of St. Louis, Missouri, for the brand Budweiser. Hollenbach was successful in introducing the beer to all the better hotels and saloons of Louisville and the surrounding area. They had foreign connections with Germany, France, Spain, Portugal, Italy, Hungary, Holland, Denmark, Switzerland, Sweden and many other countries. They carried Moselle wine, Sect wine and Burgundy wines.[128]

When Prohibition closed the doors of the Glencoe Distillery, the company had a large reserve of stock of its whiskey in bond. As a result, the company was permitted to continue in business and was able to supply the pharmaceutical trade during those years. In 1929, when Prohibition ended and whiskey making resumed in Kentucky, the Glencoe Distillery Company participated with several other distilleries in rehabilitating one of the four remaining distilleries in Kentucky. With the repeal of Prohibition, Glencoe was one of the few distilleries in the United States that had a stock of two-, three- and four-year-old whiskies on hand and was able to meet the new demand for bourbon.[129]

Phil Hollenbach tombstone, Cave Hill Cemetery. *Photo by author.*

On March 21, 1929, at the age of seventy-seven, Philip J. Hollenbach passed away at Saints Mary and Elizabeth Hospital. The funeral was held at his son Louis J. Hollenbach Sr.'s home at 1805 Windsor Place, and he was laid to rest at Cave Hill Cemetery, Section 4, Lot 17, Grave 6.

In 1935, Glencoe built a new distillery on Cane Run Road, and in 1939, the bottling plant was moved to the Cane Run Road location. It could produce 100 barrels in eight hours and could bottle 1,500 cases per day. Three warehouses held 60,000 barrels. In 1943, the National Distillers Product Corporation and McKesson & Robbins Inc. bought the distillery for $3 million, including 37,858 barrels. Louis J. Hollenbach Jr. was kept on as distillery superintendent.[130]

Louis J. Hollenbach Sr. died on October 15, 1954, at his home in the Commodore Apartments. He was seventy-three years old. He bred and raced Thoroughbred horses and had a stable of twenty-three racing horses on the Kentucky and Chicago circuit. He owned Golden Maxim Stable Inc. He was buried at Calvary Cemetery.[131]

CHAPTER 10

JAMES THOMPSON, FRANK THOMPSON, FRANK THOMPSON JR. AND JAMES "BUDDY" THOMPSON

Glenmore Distillery

James Thompson was born in Longfield, Ireland, not far from Londonderry, on May 5, 1855. When he was fifteen years old, he moved to America and found employment with his uncle, John Getty of Bowling Green, Kentucky, in the dry goods business. He left Bowling Green, moved to Louisville and found employment in the dry goods store of Knott & Sons. By 1876, he had quit the dry goods store and found employment as a salesman in the wholesale liquor company of Chambers & Brown. He was a cousin of the company's owner, George Garvin Brown. In 1881, Chambers retired, and Thompson became a member of the company. The company changed its name to Brown-Thompson & Company.[132]

In 1889, Thompson sold his share of Brown-Thompson & Company to George Forman in order to create his own wholesale liquor company. In 1905, James, along with his brother, Francis P. Thompson, incorporated the company, with his brother becoming a partner. James Brothers & Company remained a wholesale liquor company. In 1891, Francis died. In 1900, James bought the vacated building owned by N.M. Uri & Company on 1900 First and Main Streets on "Whiskey Row."[133]

In 1901, James Thompson bought the Glenmore Distillery, which was located in Owensboro, Kentucky. Glenmore was established by Richard Monarch in 1872, and in 1898, the distillery went into bankruptcy; Thompson bought the distillery at auction for $30,000. Once Thompson acquired Glenmore, he became president of the company and placed his brother-in-law, Harry S. Barton, in charge of the distillery. Barton remodeled the distillery and made it into one of the largest distilleries in the nation.[134] Since James had acquired a distillery, he changed the company's name to Jas. Thompson & Brothers, Distillers and Whiskey Merchants. He was able to market several brands of bourbon, including Kentucky Tavern, Old Thompson and Jefferson County Bourbon.

Since the 1919 passage of the Volstead Act, also known as the National Prohibition Act, Glenmore was only one of seven distilleries permitted to make medicinal whiskey during Prohibition.

James Thompson was a student of the political, economic and industrial situation in Ireland and wrote numerous articles on the subjects. He was also interested in the hospital societies of Ireland and supported the Londonderry City Hospital. He was an advocate of good roads and had served in public capacities, although he never held political office. He served as assistant federal food administrator to Kentucky, under Senator-elect Fred Sackett.[135]

Thompson also owned large industrial and agricultural interests in Davies County and was the only man living in the state of Kentucky who had

Frank Thompson tombstone, Cave Hill Cemetery. *Photo by author.*

James Thompson tombstone, Cave Hill Cemetery. *Photo by author.*

rejected a baronetcy from the British king. Thompson declined the offer with thanks and remarked, "I would rather be a plain citizen of Kentucky than an Earl of the Empire." He was the author of several brochures on the settlement of his family in Kentucky and was proud of his citizenship.[136]

Thompson's health had been failing for years, and on December 11, 1924, he died at his home in Anchorage, Kentucky. His funeral services were held in his home, and he was laid to rest at Cave Hill Cemetery, Section G, Lot 72, Grave 9. He was survived by his sons James P. Thompson and Frank B. Thompson; a granddaughter, Helen Thompson; and four brothers: Dr. Cuthbert Thompson, Reverend George Thompson, Samuel Thompson and John Thompson.[137]

In his will, he left his widow, Attie Thompson, $335,000, including his home and all the contents, as well as a residence in Osbrey, Florida. The rest of his estate was to be divided between his widow and his two sons.[138] Attie Thompson left $274,000 to her six grandchildren.

In 1924, Frank Barton Thompson became president of Glenmore Distillery. Frank was born in Anchorage, Kentucky, on July 4, 1895. He graduated from Louisville Male High School and the University of

Michigan, where he studied engineering. In the spring of 1917, he enlisted in the army, entered the Officers' Training Corps and spent a brief training period at Fort Benjamin Harrison, Indiana. He spent eighteen months overseas with the Second Infantry Division and took part in every major campaign involving U.S. forces. By the war's end, he had earned the rank of second lieutenant and was awarded the French Croix de Guerre. In 1919, he returned to civilian life.[139]

In 1933, the Volstead Act was repealed, and Glenmore had more than thirty-one thousand barrels of aged whiskey ready to be bottled and sold. Frank Thompson and his company went back to producing bourbon on a large scale.

When Pearl Harbor was attacked on December 5, 1941, Frank Thompson's commission as a captain in the Kentucky National Guard was activated; however, he decided to enlist as a private in the regular army. He turned over the presidency of Glenmore to Owensboro native Joe Engelhard. During the next four years, Thompson rose to the rank of lieutenant colonel and became battalion commander of the 158th Regimental Combat Team, called the "Bushmasters." The Bushmasters' missions often included pre-invasion landings and reconnaissance. During World War II, he took part in campaigns in New Guinea, the Bismarck Archipelago and the Philippines.[140]

In April 1944, while Thompson was in the Pacific, Engelhard bought the Taylor & Williams distillery for $5 million to $6 million. The Taylor & Williams distillery made six hundred bushels per day and sixty barrels of bourbon per day. The still, the sixty-thousand barrel warehouse and a bottling house with a capacity of 2,400 cases daily were located at Seventh Street Road. They made Kentucky Tavern, Old Thompson, Glenmore, Tom Hardy and Mint Springs. The company survived through fire and flood, and in 1946, it made its 2 millionth bottle of bourbon.[141]

After World War II, Frank Thompson joined the Army Reserve after leaving the regular army and retired with the rank of colonel in 1955. In 1947, he was awarded the Novelli Award for Distinguished Service, which was previously given to General George Marshall and General Joseph Stillwell by New York's Franklin Roosevelt American Legion Post. During the 1950s, Frank became a living symbol of Glenmore's Old Kentucky Tavern label, wearing a Texas-style hat and high leather shoes in the advertisements. After he returned from World War II, he became chairman of the board for Glenmore.[142]

His son Frank Thompson Jr. was born in 1933. In 1955, he became a sales trainee after graduating from Yale University. He was also a lieutenant in the U.S. Air Force. He became vice-president for sales and a director for two years and eventually executive vice-president of the company. In 1964, Thompson was appointed president of the company at the age of thirty-one. He was one of the youngest presidents of any major company in the United States. He made many of the key decisions involved in Glenmore's two major acquisitions: the purchases of the Mr. Boston Distiller Corporation and Foreign Vintages Inc. The addition of Mr. Boston increased the range of the company's domestic products, while Foreign Vintages added several major foreign brands of liquor to the Glenmore line.[143]

When Frank Thompson Jr. joined the company, Glenmore's line of products consisted of three major brands. Thompson expanded that to include vodka, cordials, tequila, wine and other products. While he was president of the company for ten years, he increased sales from $68 million to $170 million annually.

In his spare time, Frank Jr. was active in gardening and forestry on his Oldham County farm. He was a member of the Louisville Country Club, the River Valley Club, the Pendennis Club and the Wynn Stay Club. He resigned as president of the distillery on October 2, 1974, when doctors discovered a malignant tumor. He died on May 13, 1975. He was only forty-two years old. He was survived by his wife, Anne McGill Hogue; two sons, Frank Thompson III and Christopher; and a daughter, Rebecca. His funeral was held at St. Francis in the Fields Episcopal Church.[144] On February 28, 1990, at the age of ninety-four, Frank Thompson Sr. died at his Louisville home. He was survived by his son James Thompson; his wife, Ida Mane Webb; his daughter, Mrs. Robert Nash; thirteen grandchildren; and six great-grandchildren. His funeral was held at Highland Presbyterian Church, and he was laid to rest at Cave Hill Cemetery, Section G, Lot 72, Grave 14.[145]

In 1974, James "Buddy" Thompson became chairman, chief executive and president of Glenmore Distillery in 1982. James "Buddy" Thompson was the brother of Frank Thompson Jr. In June 1988, Glenmore added more bourbon, gin and vodka to its portfolio and agreed to buy the rival firm of Medley Distilling Company for $7 million, acquiring more than one hundred brands. It sold off many of the brands but kept Ezra Brooks bourbon, Skol gin and vodka and Philadelphia blend whiskey. It sold its wine business. It also bought the Fleischmann Distilling Company for $70 million. The acquisitions tripled the company's sales, which amounted to $300 million and made the company the seventh-largest distilled spirits

Frank Barton Thompson grave marker, Cave Hill Cemetery. *Photo by author.*

company in the nation. The company's brands included Kentucky Tavern, Mellow Mash and Yellowstone bourbons. It also had the line of Old Thompson whiskey; the Boston brandies, cordials and cocktails; Hot Shot liqueur; Chi-Chi's margaritas; Ezra Brooks bourbon; and Skol gin and vodka. It also had Fleischmann gin and vodka, Scoresby and Inver House Scotches and Canadian Ltd. whiskey. It also marketed Icelandic vodka and Desmond and Duff twelve-year-old Scotch.[146]

In July 1991, United Distillers, owned by a subsidiary of London-based Guinness PLC, which was the world's second-largest distillery company, approached "Buddy" Thompson and made an offer to buy the company for $161 million. Those who held 90 percent of the stock and were entitled to vote on the deal approved of the buyout. James Espey, president and chief executive officer of United Distillers North America, let the Glenmore Company keep its name but become a wholly owned subsidiary. Thompson would remain chairman of Glenmore indefinitely and become chairman of United Distillers USA. He also served as a select management executive officer of Glenmore. The decision to sell the company came about because there was no one else in the Thompson family to run the company. When

United Distillers bought Glenmore, it became the third-largest distillery in the United States and pushed Brown-Forman to fifth largest. At the time, IDV/Metropolitan, a British firm, was number one, and Seagrams was ranked the second-largest distillery.[147] In 1996, the company changed its name from United Distillers Glenmore to United Distillers USA. In 2009, the Sazerac Company purchased the distillery and brought back the name Glenmore Distillery.

CHAPTER 11

NATHAN BLOCK

Kentucky Oaks, Kentucky Derby, Tremont and Gold Dust

Nathan F. Block was born in Bavaria, Germany, on August 24, 1844. He came to America when he was ten years old. In 1870, he married Clara Gotthelf. In his early business career, he was associated with the wholesale dry goods business, but in 1888, he entered the whiskey brokerage business under the name of Frankel & Block, with his offices located at 144 West Main Street.[148]

In 1875, his brother, Joseph Block, who was born on September 11, 1850, and L. Franck formed the partnership of Block, Franck & Company. They were later joined by E. Franck. Their offices were located at 205 West Main Street. The Francks were also formerly in the wholesale dry goods business. Block, Franck & Company was devoted to the handling of its Kentucky Oaks whiskey. The firm was one of the first to introduce the selling of fine Kentucky whiskies in bond in the retail trade. It was very successful in placing its well-known brands in the hands of the best class of retail dealers in all parts of the country. It sold its goods in twenty-nine states. Not only did it make Kentucky Oaks, but it also made the Kentucky Derby handmade sour mash bourbon and rye whiskies. It also added the Tremont and Gold Dust hand-made sour mash whiskies to its brands.[149]

In addition to its own brands, it also sold in bond a number of other popular brands of Kentucky goods. It was able to supply the trade with any fine whiskey made in Kentucky. The firm dealt only in goods in bond, and all shipments were made direct from bonded warehouses, ensuring the trade that the whiskies were perfectly straight, which was the most desirable

Nathan Block's advertisement. *From* The Industries of Louisville and New Albany, Indiana *(1886)*.

feature for retailers in buying their goods. Shipments were made from its distillery, located at Pleasure Ridge Park.[150] In 1891, Nathan and his brother, Joseph, were part owners with Isaac Bernheim of the Pleasure Ridge Park Distillery, which burned in 1896 and was never rebuilt.[151]

On June 1, 1903, Nathan Block retired from the whiskey brokerage business, having been in industry for twenty-three years. When he retired, the firm of Nathan and Sons was dissolved and was replaced with his sons, Joseph Block and Bernard Block, who were already partners in the firm. The name was changed to Block Brothers.[152]

In the early 1890s, Nathan was president of the Temple Adath Israel congregation. He also served as president of the Standard Club in 1909. He was past master of the St. George Lodge, Free and Accepted Masons, and also of the B'nai B'rith Lodge. He was also involved in charity work and was identified with the Hebrew Relief Society and other organizations. He was also a member of the Commercial Club and the Louisville Board of Trade.[153]

In 1907, Block sold his home on the southeast corner of Third Street for $20,000 to W.H. Curtis of Eminence, Kentucky, who planned to use the house as a winter home. Curtis was a prominent farmer of Henry County.[154] On January 1, 1917, Nathan Block died at the age of seventy-three at his home on 2017 Baringer Avenue. He had suffered from asthma and other illness for many years. He was survived by his son, Bernard Block, and three daughters: Mrs. Violet Block Goldsmith, married to Norton Goldsmith; Mrs. Elise Block Mayer, married to Max Mayer; and Mrs. Cora Block Spangenthal, married to Adolph Spangenthal. He was also survived by his brother, Joseph Block.[155] According to burial records, Joseph died in 1939 and was buried at Adath Israel Cemetery. He also had four sisters who survived him: Mrs. Dora Block Weil, married to Leopold Weil; Mrs. William Fleishaker; Mrs. C.K. Weil; and Mrs. Clara Lazarus.

When he died, Nathan Block was worth $100,000. He was personally worth $30,000, and his realty was valued at $10,000 in Louisville and $60,000 of property in Buffalo, New York. After the cash payment of $1,900 and the setting aside of a trust of $3,000 for the education of Norton Goldsmith, a grandson, the remaining estate was to be divided between Violet Goldsmith, Cora Spangenthal, Elise Mayer, Bernard Block and Ethel and Nathan Block Jr.—children of Joseph Block, his son. He also left money to the Young Men's Hebrew Association, the United Hebrew Association of Louisville, the Jewish Orphan Asylum of Cleveland, the Jewish Hospital Association and the Hebrew Ladies Sewing Circle.[156]

On January 25, 1908, Joseph committed suicide by firing a bullet into his brain while standing before a mirror in the toilet adjoining his office. He had been grieving over the death of his mother, who died the previous year. He had also dealt with depression. For about five months before his suicide, he had been suffering and was being treated in a mental asylum in Philadelphia. He went to his office as usual and came to town in a car with John C. Weller, of the firm John C. Weller & Company. He gave no indication of what he was going to do. His father, Nathan, requested that his son accompany him to the ticket office. His father was planning to go on a trip to Florida, but his son declined the offer, stating that he wanted to read some mail. Shortly before he committed suicide, Joseph made several trips to the toilet. Soon the shot was heard, and Nellie West, the stenographer, and Lula Garth, the bookkeeper, ran to the barbershop of Proctor Brothers at 148 West Main Street, two doors away, to asked for help. W.E. Proctor and Martin Evans, who were in the shop, ran to the scene. They managed to find a ladder and climbed through the skylight of the toilet room door, as the door was locked. They found Block lying dead on the floor. A .38-caliber revolver lay only a few inches away from his left side; the bullet had entered the right temple. He was only thirty-five years old. He was survived by his wife, Cora Rosenau, and two children, Nathan Jr. and Ethel. He was buried at Adath Israel Cemetery.[157]

After the death of Joseph, Bernard continued to run Block Brothers until Prohibition. He later became the founder of Block, Fetter & Trost, stockbrokers, which later became Stein Brothers & Boyce Inc. He died on September 3, 1966, at the age of eighty-four. He is buried at Adath Israel Cemetery. He was survived by his wife, Eidee Weiss; a daughter, Lucy Block, who married U.S. Navy ensign Monroe Heumann Jr.; and three grandchildren.[158]

Joseph was not the only son of Nathan Block to commit suicide by bullet. His son Walter Block had been in ill health for several years. He had spent most of his time at health resorts. On May 21, 1915, he took a gun and shot himself at the home of Nathan Block at 2017 Baringer Avenue. He was also standing in front of a mirror when he committed suicide. He was also buried at Adath Israel Cemetery.[159]

CHAPTER 12

GEORGE SWEARINGEN

The Mellwood Distillery

George Swearingen was born on March 12, 1838. He came from nine generations of Swearingens. Gerritt Van Swearingen was the first family member to settle in America. He was one of the younger sons of a Dutch nobleman and a native of Beemsterdam, North Holland. In 1856, he was sent to America in command of a ship full of supplies for the Dutch colony at New Amsterdam, which later became New York. The ship was lost in a storm off the Atlantic coast, which led to Captain Van Swearingen to abandon the sea and settle in Maryland. Four generations of the family lived in Maryland. Toward the end of the seventeenth century, the "Van" was dropped from the name. In 1804, some of the family moved to Bullitt County, Kentucky. George Swearingen's father, William Wallace Swearingen, was born in 1803; they moved from Maryland and settled in Bullitt County, Kentucky, in 1804. William became successful farmer. He married Julia Crist, who was the daughter of Henry Crist, one of Kentucky's distinguished pioneers. In 1780, Henry Crist fought a bloody battle at the Salt River with the Indians. He was also a frontier farmer, and from 1795 to 1796, he served as a member of the Kentucky legislature. In 1808, he was elected to Congress and represented Louisville from 1809 to 1811. Julia Crist died in 1838, not too long after George was born. William Wallace Swearingen lived until 1869.[160]

George attended Mount Washington Academy, and at the age of sixteen, he entered Centre College at Danville, Kentucky, graduating in 1857. He taught for about a year, and after his graduation, he married Mary Embry.

In 1860, his father, William, sold George the family farm in Bullitt County. George also ran a small distillery on his farm that he called Millwood. In 1866, George gave up his career as a farmer and moved to Louisville.[161]

After a brief period working in the grocery business, in 1869, George built and ran the Mellwood Distillery. George originally wanted to call his bourbon Millwood, but due to error of the printer, the brand name was misspelled as Mellwood; Swearingen decided to keep the name. He began his distillery on a small scale, but it soon became one of the largest and most successful distilleries in the state. It was located at Mellwood (formerly Reservoir Road) and Frankfort Avenues and was considered at the time one of the most complete and best-conducted establishments of its kind in the world. The plant occupied twelve acres and had five immense bonded warehouses and one "free" warehouse. The distillery had an equipped mill room with a series of patent roller mills. It also had a large fermenting house that contained nineteen vats or tanks, with each having a capacity of ten thousand gallons, besides the vast beer well in the center. The plant also had a boiler room, barrel rooms, a branding house and large cattle feeding pens with accommodations for one thousand head of cattle. For four months of the year, the distillery would shut down. One of its warehouses could hold thirty-six thousand barrels of bourbon. The warehouses were steam heated, and an even temperature was maintained in all parts of the building.[162]

By 1909, the distillery had become well known in the trade; sales were large, and the high-grade quality of whiskey was established. About ten thousand barrels and fifty thousand cases were sold annually and manufactured at the cost of $200,000; the distillery was worth $600,000. The brand trademark whiskies and bourbons made by the Mellwood Distillery included Runnymede Club Whiskey, Mellwood Whiskey, Runnymede Club Pure Rye, Runnymede Club Bourbon, Normandy Pure Rye Whiskey and Old Water Mill Whiskey. It also made G.W.S. hand-made sour mash and Dundee fire copper bourbon.[163]

In 1905, Swearingen and the Mellwood Distillery Company sued Samuel R. Harper, Cyrus F. Reynolds and the citizens and residents of the Fort Smith division of the Western District of Arkansas under the name of Harper-Reynolds Liquor Company for infringement. Mellwood insisted that they were the sole and exclusive owner of the Mellwood whiskey name. Swearingen stated that his trademark Mellwood bourbon was worth $500,000. Swearingen alleged that Harper-Reynolds sold its whiskey under the trade name Mellwood and fraudulently intended to "invade complainants good will and divert its trade, made, or caused to be made, or

MELLWOOD DISTILLERY CO.,
LOUISVILLE, KY.,
DISTILLERS AND PROPRIETORS OF
MELLWOOD,
Fire Copper Bourbon.
"G. W. S.,"
Hand-Made Sour Mash.
Fine Kentucky Whiskies.
NORMANDY,
Pure Rye.
DUNDEE,
Fire Copper Bourbon.

Mellwood advertisement. *From* Wine and Spirit Bulletin *(1903).*

sold and caused to be sold in the city of Ft. Smith, Arkansas and elsewhere, a certain spurious liquor, not produced by said complainant, to which they have caused the trademark 'Mill Wood' to be affixed…the defendants have diverted to themselves the trade of complainant and have invaded its good will." They also stated that Harper-Reynolds had labels that stated its bourbon was made by the "Kentucky Mill Wood Distilling Company." Swearingen stated that the Harper-Reynolds company used the name solely for fraud and deceit in order to pass off its bourbon as Swearingen's. In 1909, the courts granted an injunction restraining Harper-Reynolds from the use of the label, and the case was referred to the master to ascertain the profits and damages, as well as for the cost of the suit.[164] The judge stated that there was no "Mill Wood Distillery" and that the company put labels on cheap bottles of whiskey. The court stated that it would not tolerate a deception devised to delude the consuming public.

After amassing a large fortune, Swearingen sold his interest in the distillery to the Kentucky Distilleries and Warehouse Company, also known as the Whiskey Trust, and entered a different career. Along with several other men, he formed the Kentucky Title Company and became president of the organization. Not too long after, he became a banker. In 1890, he organized the Union National Bank, located at Sixth and Main Streets in Louisville. The Union National Bank became one of the strongest and most popular banks in the city of Louisville.[165] Swearingen was connected prominently to the Presbyterian Church, and for years, he was president of the board

Union National Bank. *University of Louisville Digital Archives, R.G. Potter Collection.*

of trustees of the Presbyterian Orphanage of Louisville and contributed generously toward the care and maintenance of the wards of the church.[166]

In 1898, Swearingen suffered a stroke that led to temporary paralysis. He continued to have strokes and paralysis. After suffering another stroke and after being paralyzed for several weeks, he died on December 18, 1901, at

George Swearingen monument, Cave Hill Cemetery. *Photo by author.*

George Swearingen marker, Cave Hill Cemetery. *Photo by author.*

his home at 218 West Broadway. His funeral was held at his home, and his pallbearers were a who's who of Louisville dignitaries. His pallbearers were R.S. Veech, L.O. Cox, Thomas Bullitt, John H. Leathers, H.C. Rhodes, W.F. Booker, William Hamilton and R.F. Balke. His honorary pallbearers were William Robinson, R.T. Durrett, W.C. Priest, Dr. J.W. Aiken, Walter N. Haldeman, John Doerhoefer, Bennett Young, Samuel Avery, C.G. Strater, J.K. Lemon, W.S. McRae, Fred Hoertz, J.G. Schmidlapp, Dr. J.G. Cecil, Dr. John A. Ouchterlony, Hancock Taylor, George Brown, R.W. Delph, Charles Lindenberger, Alfred Pirtle and Judge W.B. Hoke.[167] He was buried at Cave Hill Cemetery, Section A, Lot 201, Grave 7. On his tombstone are these words: "A friend of truth, of soul sincere, In action faithful, and in honor clear."

The Mellwood Distillery continued to operate under the "Whiskey Trust" until Prohibition closed its doors. After the repeal of Prohibition in 1934, the distillery was renovated and put back into production. It was at that time owned by the General Distillers of Kentucky Corporation, which made Kentucky Nectar, Lick Run and Old Chuck. In 1974, the Mellwood Distillery finally went out of business.

CHAPTER 13

HENRY McKENNA

McKenna Distillery

Henry McKenna was born in County Derry, Ireland, on January 9, 1819. He moved to America when he was twenty years old. He worked in Central Kentucky as a turnpike builder and settled in Fairfield after building the road from Fairfield to Bloomfield. In 1855, he built his distillery, having learned distilling in Ireland. He began to manufacture whiskey on a small scale in an old shed, making less than one barrel per day. He made sure that his whiskey was of a superior quality, and he soon discontinued his flour mill to focus on the distillery.[168]

When the Civil War ended, he increased the capacity of his distillery to a little more than one barrel per day. In 1883, he built a new brick distillery, increasing the capacity to three barrels per day, while still maintaining the quality. The method he used to make his McKenna Sour Mash Whiskey was described in 1896 by Sam Carpenter Elliott in the *Nelson County Record*, reprinted in the *Kentucky Standard*:

> [T]*he corn used is first thoroughly examined. The tip ends, where rotten grain is most to be found, is examined, and these diseased grains are shelled off by hand. When this is done, the corn is put into a sheller, and afterwards thoroughly fanned. So is the malt and rye. By this means there is nothing allowed to be put into the mash but pure grain. Everything around the premises denotes cleanliness and clock work management. The small, old fashioned mash tubs are placed upon a contrivance and after attaching*

a rope to it are pulled up in position. While one man pours water into the tub through a rubber hose, another stirs the meal and draws it down into the opening, which carries the mash into a stirring machine, where it is finally conveyed through wooden channels into fermenting tubs to undergo the required fermenting period before ready for use. The warehouses are built upon a rock foundation and are well ventilated and conveniently arranged. They have a capacity of eight thousand barrels. There is one thing that can be said of another in Kentucky. That is, they never sell their whiskey in bond until it is more than three years old.[169]

McKenna's plant was located eleven miles northeast of Bardstown and four miles west of Bloomfield, on State Highway 48. He also owned and ran a five-hundred-acre farm that grew the grains he used in his distillery. Henry McKenna gave his whiskey a worldwide reputation. If he heard of a dealer diluting his whiskey or selling another whiskey as the McKenna brand when his supply ran out, he would not do business with that dealer again.[170]

McKenna was a devout Catholic and was noted for his liberality. Whenever he could aid the needy, he did not hesitate. He left his distillery to his sons and stipulated that the name should always remain McKenna. His home was the finest residence in Nelson County, and he was also the wealthiest man in the county. His son Daniel McKenna was in charge of the distillery, his son James McKenna conducted the wholesale business and his son Stafford McKenna was the traveling salesman. James lived at Sixth and Kentucky Streets in Louisville, with his office at 245 Fourth Street. His only daughter, Mary McKenna, lived with her father and attended to the upkeep of the house. On February 22, 1893, Henry died at his home at Fairfield, Nelson County. His wife, Elizabeth, had died several years earlier in 1880.[171]

Henry McKenna. *From* Pharmaceutical Era *9 (January 1–June 15, 1893, and March 15, 1893), edited by Charles Parsons.*

Before Prohibition, his sons James and Stafford McKenna had operated the distillery for twenty-five years. Coleman Bixler was in charge of distillation for sixteen years before Prohibition. During Prohibition, McKenna Distillery, along with a Louisville distillery, was one of the few plants in Kentucky that was not dismantled and

Henry McKenna Bourbon advertising, bourbon jug. *Ryan Reynolds Collection.*

was allowed to make twenty barrels per day for medicinal purposes. In 1934, operations resumed at the McKenna plant, and James McKenna became president of the distillery. Stafford McKenna died on January 16, 1935, at Saint Joseph's Infirmary, where he had underwent an appendicitis operation. His funeral was held at his residence in Fairfield, and he was buried at St. Michael's Church cemetery. He had six daughters.[172] James died on December 17, 1941, at the home of his son-in-law, Dr. M.J. Henry, on 1226 Summit Avenue in Louisville. His wife was Mary Ellen White McKenna. His funeral was held at St. Michael's Church in Fairfield, and he was buried at St. Louis Cemetery in Louisville.[173]

In 1941, the distillers corporation Seagrams bought Henry McKenna Inc. for $950,000. James McKenna, former president; Dr. Henry J. McKenna, former executive vice-president; and their associates sold all of their stock.[174] Heaven Hill Distillery, in Bardstown, Kentucky, continues to make Henry McKenna bourbon to this today.

CHAPTER 14

ISAAC AND BERNARD BERNHEIM

Bernheim Distillery

Isaac Bernheim was born on November 4, 1848, in Schmieheim, Baden, Germany. His brother, Bernard, was born on December 13, 1850. Their father was Leon Bernheim, who was born on September 19, 1808. Isaac was a merchant and the first Jewish person to open a regular store in his native town. He later became a wholesale buyer and seller of wine as well as a landowner. His wife was Fanny Bernheim. Isaac and Bernard's father died on January 9, 1856, and in 1857, their mother remarried to Louis Weil. At the age of ten, Isaac went to grade school in Ettenheim. In 1861, he became an apprentice to a commercial house for three years. At the age of sixteen, he was a clerk at a Mannheim firm, and later he gained a position with Gebrueder Elkan, which was located in Frankfort on Main, a city in Germany. Gebrueder Elkan was a wholesale dealer in knit goods. His uncle, Benjamin Weille, was in New York manufacturing cotton knitting yarn and promised to give Isaac a job in his firm if he left for America. In 1867, Isaac emigrated and landed in New York.[175]

Isaac started his career as a peddler. His uncle moved from New York to Paducah, Kentucky. On May 5, 1868, Isaac left for Paducah and became a bookkeeper and second salesman in his uncle's store. While in Paducah, Isaac met Moses Bloom. Bloom was associated with Rueben Loeb, who was in the wholesale liquor business under the firm of Loeb and Bloom. Bernheim quit his uncle's store and became a bookkeeper for Loeb and Bloom. Isaac advised his brother, Bernard, to leave Germany and become a bookkeeper at Loeb and Bloom as well. At the time, Bernard was a clerk in the office of a lawyer in Freiburg. In 1870, Bernard came to America. Isaac became a traveling salesman for Loeb and Bloom, and Bernard became a bookkeeper for the firm.[176]

On January 1, 1872, Isaac decided to start his own wholesale liquor business and quit Loeb and Bloom. Elbridge Palmer invested in the Bernheim business and became a silent partner. In 1875, the brothers bought out Palmer's interest in the firm and added Nathan Uri to the company as a partner. Isaac married Uri's sister, Amanda, in 1874. The firm changed its name to Bernheim Brothers & Uri, with Isaac as president and Uri as vice-president. The firm acted as a rectifier, buying bulk whiskey and bottled its own brand names. In 1879, Uri and the Bernheim brothers trademarked their whiskey I.W. Harper, named after Isaac Wolfe Bernheim's initials. The name Harper came from the horse breeder F.B. Harper, who had horses in the first three Kentucky Derbys. They also branded Old Continental. Both of their bourbons were considered to be equal of the finest sour mash of the times before the Civil War. They boasted that they made their whiskies exactly as Kentucky whiskies were made in the olden times, when two or three gallons were all anyone expected to get from a bushel of grain. They employed twenty-two salesmen, who covered the most of the North and were expanding into the southern parts of the country.[177] On April 1, 1888, the Bernheims decided to look for a wider field of operation and moved to Louisville. Louisville had become a major distribution center for whiskey throughout the country, mainly because of the Louisville and Nashville Railroad, which had connections to the East, the North and to southern cities as well as the Ohio River. They moved to a six-story building, 32 feet by 110 feet, at 135–37 Main Street. The company carried in stock nearly all of the celebrated Kentucky bourbons and ryes.[178]

In 1888, Bernheim and Uri bought the Pleasure Ridge Park Distillery and renamed the company Bernheim Distillery Company. The distillery could produce ten thousand barrels of sour mash bourbon and rye whiskey every year. The Park Ridge Distillery was established in 1881 by F.G. Paine & Company, near Mill Creek in the southwestern part of Jefferson County. In 1893, Nathan Uri

Bernard Bernheim. *From Ben La Bree's* Notable Men of Kentucky at the Beginning of the 20th Century *(1902).*

decided to withdraw from the firm and go into business for himself. By 1895, the Bernheims' business had expanded, and they bought the adjoining store. In 1896, the bonded warehouses of the distillery plant at Pleasure Ridge Park, owned by Nathan Block and the Bernheims, caught fire and was destroyed. Insurance covered the loss. They still owed government taxes on the whiskey that was destroyed in the warehouses, which amounted to $1 million. Lyman Gage, secretary of the treasury, canceled the bond in 1897 so the brothers did not owe the taxes.[179]

The Bernheims wanted their own distillery, and in 1896, they began construction on their plant near the city limits on the Illinois Central Railroad. In 1897, they produced their first sour mash. In 1888, the volume of business amounted to $350,000 per year. By 1902, the annual volume had grown to millions, and the capital invested amounted to several million dollars. They sold their old store to W.L. Weller & Company and bought the Bamberger, Bloom & Company store located near the Louisville Hotel, on Main Street. Bamberger, Bloom was a dry goods firm that went bankrupt in the Panic of 1893.[180]

Barney Dreyfuss, who was the Bernheims' first cousin, immigrated to Paducah from Baden in 1883. He became their assistant bookkeeper, head bookkeeper and credit man. By 1890, he had an interest in the firm, but health forced him out of the company. Bernard Bernheim married Rosa Dreyfuss, who was the eldest daughter of their uncle, Samuel Dreyfuss. They had five children. The Flarsheim brothers from Newark, New Jersey, joined the firm; they were also in the wholesale liquor business in St. Paul, Minnesota. Alfred Flarsheim was with the company until 1896, when he opened an office in New York, until he was brought back to Louisville when Dreyfuss became ill. In 1892, Morris Flarsheim became part of the firm.

Leon Solomon Bernheim, who was born on October 10, 1875, was selected to become successor to the firm. He had a common school education, and when he was fifteen, he was made an office boy in the firm. Leon resigned from the firm in 1893 and entered drama school in New York. He rejoined the firm, and Isaac made him assistant manager and put him in charge of the New York office. He became partner in the firm at the age of twenty-five. In 1905, he quit the firm.[181]

Morris Bernheim was born on July 16, 1877. At the age of nineteen, he entered Yale University. In 1900, he was made a partner in the firm. He left the firm in June 1903. His son, Elbridge Palmer Bernheim, was born on August 9, 1881, and in 1902, he graduated from Johns Hopkins University. He took Morris's position in the company.[182]

Bernheim advertisement, 1903. *From* Wine and Spirit Bulletin *(1903).*

Bertram Moses Bernheim was born on February 15, 1879, and graduated from Johns Hopkins University as a doctor of medicine.[183]

In 1902, Isaac and Bernard established a corporation under the name of Bernheim Distillery Company. The company's invested capital was $2 million. Isaac was president and Bernard vice-president. They formed the Commercial Distilling Company of Terre Haute, Indiana, in order to obtain liquor at manufacturers' cost, and in 1906, they purchased the Warwick Distillery at Silver Creek in Madison County, Indiana. They also bought a small interest in the rye distillery of Baltimore Distillery Company in Baltimore, Maryland.[184]

In 1915, Isaac retired from the distillery business, and in the 1920s, he left his home on South Third Street and his summer home in Anchorage, Kentucky, for Denver, Colorado, and later Santa Monica, California. With his millions of dollars, Isaac decided to pursue philanthropy. He donated the Thomas Jefferson monument at a cost of $50,000 and endowed $10,000 to maintain the monument, which is located currently in front of the old Jefferson County Courthouse. The monument was designed by Moses Ezekiel.[185] In 1896, he founded the Young Men's Hebrew Association and donated a new building on the east side of First Street, south of Walnut. Isaac organized the association in 1890 and was the first president of the organization. Clark and Loomis designed the new building.[186]

In 1927, Bernheim donated the statues of Ephraim McDowell and Henry Clay to the statuary hall at the Capitol in Washington, D.C. The statues were designed by Charles Henry Niehaus.[187]

On May 10, 1929, Bernheim incorporated the Isaac W. Bernheim Foundation and purchased more than thirteen thousand acres of Kentucky Knob land situated twenty-five miles from Louisville, an estate that exceeded twenty square miles.[188] He also donated the Abraham Lincoln memorial in 1922 in front of the Main Branch of the Louisville Free Public Library. The statue was designed by George Gray Bernard.[189] Bernheim also donated the library building of the Hebrew Union College in Cincinnati, Ohio, and donated the Orphan Society and Old Men's Home in Germany.

On November 4, 1926, Bernheim donated the *Let There Be Light* statue for his wife at Cave Hill Cemetery. The statue was designed by George Gray Bernard. His wife had died on December 9, 1922. On November 14, 1923, Leon Bernheim passed on. On July 27, 1925, Bernard Bernheim died. On May 8, 1926, Morris Bernheim died.[190]

On April 1, 1945, Isaac Bernheim fell to his death from his eighth-floor apartment in Santa Monica, California. He left three daughters—Millie Rauh, Maguerite Block and Helen Roth—and one son, Bertram Bernheim. Police ruled his death a suicide. He was ninety-six years old. In his will, he left his $3 million to the Isaac W. Bernheim Trust for the Bernheim forest.[191] He was originally buried with his wife at Cave Hill Cemetery, but in 1956, his remains, along with his wife's and the statue, were moved to Bernheim Arboretum and Research Forest.

Just before Prohibition, both the Bernheim brothers and the Warwick plants were partially dismantled and the property sold, except the Bernheim Distilling Company, which made medicinal whiskey. After the end of Prohibition, in 1934, the Canadian Schenley Distillers bought the I.W. Harper brand, including the Old Charter, Belmont and Astor brands of rye and bourbon. Frank Bernheim—son of Bernard Bernheim, former vice-president of Bernheim Distillery and the last Bernheim in the whiskey business—died in New York on August 23, 1962. He lived at the Weissinger-Gaulbert apartments. He had just returned from a vacation ocean cruise and had stopped off in New York before his return to Louisville. He collapsed in a hotel. He was never married and left $6 million to his sister, Mrs. Malvin Haas.[192]

Eventually, Schenley sold the company to United Distillers Incorporated, which was owned by Guinness PLC. In 1999, Heaven Hill Distillery bought the Bernheim Distillery and refurbished it to become Heaven Hill's main distillery plant. In 2005, Heaven Hill Distillery introduced the Bernheim Original wheat whiskey brand. United Distillers bought the I.W. Harper brand and sold it exclusively for the Japanese market. In 1997, United Distillers became Diageo. In 2014, Diageo stated that I.W. Harper brand would be reintroduced into the American market and that the bourbon would be aged at the Stitzel-Weller Distillery in the southwest area of Louisville and bottled at its George Dickel plant in Tullahoma, Tennessee.

Today, the legacy of Isaac Bernheim can still be seen in Jefferson and Bullitt Counties. You can visit the graves of Bernard Bernheim and his wife and the monument *Let There Be Light* at the Bernheim Arboretum and Research Forest. While at the arboretum, you can see the many walking trails and visit the new visitor's center and restaurant. You can gaze at the Abraham Lincoln statue in front of the Louisville Main Public Library and the Thomas Jefferson statue in front of the old Jefferson County Courthouse; you can also see the Henry Clay and Ephraim McDowell statues in Washington, D.C., at the National Gallery.

CHAPTER 15

MARION TAYLOR

Old Charter Distillery

Marion Taylor was born in Greensburg, Louisiana, on July 20, 1853, and was the son of Dr. Augustin B. Taylor and Elizabeth Taylor. His father was attorney general during the Mexican-American War. He was educated in parochial schools in Louisiana and Lane University, Le Compton, Kansas.[193]

His first business venture was located in Natchez, Mississippi, where he was in the mercantile business for eight years. In 1884, Taylor moved to Louisville, and in 1886, he organized the firm of Wright & Taylor, with John J. Wright as his business partner. Wright had been in the wholesale liquor business for twenty-five years and was one of the best liquor experts in the South. Wright & Taylor was in the wholesale liquor business. According to the *Louisville Courier-Journal*, Taylor had been in the brokerage business, representing some of the leading manufacturers in the country for the southern states, with headquarters in St. Louis. Wright & Taylor occupied a three-story building on 136 Third Street, had a large trade throughout the United States and employed six traveling businessmen. It specialized in fine Kentucky whiskies, free or in bond, and made its own rectified or blended specialties, including Pride of Louisville bourbon, Pride of Louisville pure rye, Cane Spring pure rye and Fine Old Kentucky Taylor whiskey.[194]

In 1892, it bought the Old Charter Distillery, Reg. No. 266, Fifth District Kentucky, Chapeze Station, in Nelson County, Kentucky. The distillery was built in 1874 by A.B. Chapeze. The distillery had been in existence for thirty

Marion Taylor. *From Ben La Bree's* Notable Men of Kentucky at the Beginning of the 20th Century *(1902).*

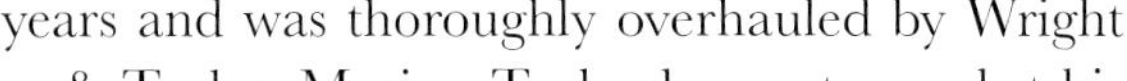

years and was thoroughly overhauled by Wright & Taylor. Marion Taylor began to market his own straight bourbon brand, named after the Old Charter Distillery.[195]

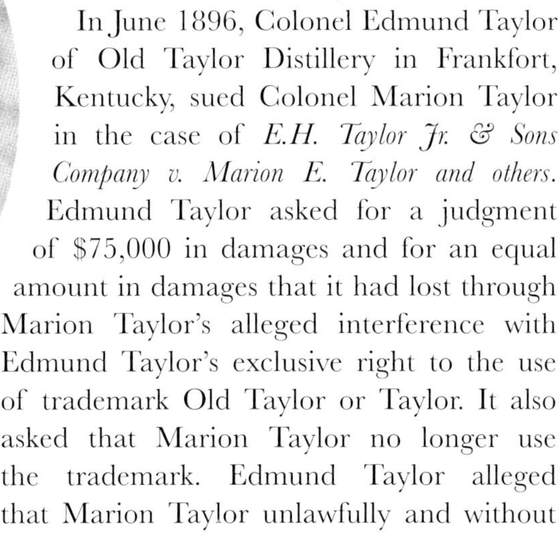

In June 1896, Colonel Edmund Taylor of Old Taylor Distillery in Frankfort, Kentucky, sued Colonel Marion Taylor in the case of *E.H. Taylor Jr. & Sons Company v. Marion E. Taylor and others.* Edmund Taylor asked for a judgment of $75,000 in damages and for an equal amount in damages that it had lost through Marion Taylor's alleged interference with Edmund Taylor's exclusive right to the use of trademark Old Taylor or Taylor. It also asked that Marion Taylor no longer use the trademark. Edmund Taylor alleged that Marion Taylor unlawfully and without Edmund Taylor's consent used on packages, barrels and bottles, containing spurious compounds of whiskey, a trademark and brand substantially the same and almost identical to that of Edmund Taylor. Edmund Taylor alleged that Marion Taylor used the words "Taylor" and "Old Taylor" unlawfully and fraudulently in connection with the words "Fine," "Ky." and "whiskey" on bottles and packages and barrels containing whiskey and that the words used were calculated to mislead the public, purchasers and consumers of whiskey into the belief that said bottles, barrels and packages branded, manufactured and owned by Marion Taylor were actually products of Edmund Taylor. Edmund Taylor wanted the courts to restrain Marion Taylor from infringing on its trademark "Old Taylor." Edmund Taylor also stated that there was no Wright and that Wright & Taylor did not exist. He also stated that Marion Taylor was "only a jobber and blender of other people's whisky; that he is not a distiller, nor the owner of a distillery, nor has he ever owned any brand containing the words "Old," "Fine," "Ky." or "Old Taylor."[196] On June 10, 1903, the Jefferson Chancery Court dismissed the claim for damages, therefore reserving the right to sue in another action if Marion Taylor should desire.[197] Immediately after the decision, Marion Taylor took out an ad in *Wine and Spirit Bulletin* publishing the Jefferson County Chancery Court's decision:

Left, top: Wright & Taylor advertisement, 1903. *From* Wine and Spirit Bulletin *(1903).*

Left, bottom: Old Taylor Distillery Company bourbon decanter. *Ryan Reynolds Collection.*

Right: Old Charter advertising ribbon. *Author's collection.*

> *The defendant, Marion E. Taylor (Wright & Taylor) in branding his bottles, cases, packages, and barrels with the trademark "Taylor" or "Old Taylor" or "Kentucky Taylor" or "Fine Old Kentucky Taylor" had a perfect right to so brand them. The defendant has sustained each and every defense interposed as a bar to this suit....Second, that priority of adoption and use by others of the name "Taylor" and "Old Taylor" precludes the plaintiff from claiming the said words as a trademark; the plaintiff has failed to sustain its suit and its petition must be and is dismissed with costs.*[198]

Edmund Taylor decided to appeal the decision made by the lower court. On March 17, 1905, Chief Justice Tolbert Frank Hobson of the Court of Appeals reversed the judgment of the Jefferson Chancery Court. In reversing the judgment, Marion Taylor was allowed to properly sell his brand of Old Kentucky Taylor, provided he framed his advertisements to show that his was a blended whiskey, but he was not allowed to sell his blended whiskey under labels or advertisements that concealed the true character of his whiskey, "for this would destroy the value of E.H. Taylor's trade."[199]

In the action of fraudulent reproduction of the plaintiff's goods, the court stated that "there could be no accounting of profits in equity. That the remedy was by the common law action for damages as in any other case of fraud. So much of the action as sought damages having been dismissed without prejudice, the only remedy to which the court found appellant entitled was an injunction."[200] Marion Taylor did not have to pay any damages and was allowed to use his brand name. The courts ruled that there was a difference between blended whiskey and straight bourbon. Rectified or blended whiskey was known in the trade as "single stamp whiskey," while bonded whiskey was known as "double stamped whiskey." Rectifiers or blenders take a barrel of whiskey and then draw off a large part of it, filling it with water and adding spirits or other chemicals to make it proof and give it age, bead and so on. Blended whiskey was a cheaper whiskey and was a temptation to simulate the more expensive straight whiskey. The courts stated that Marion Taylor intentionally misled consumers through his advertising by trying to pass off his blended Fine Old Kentucky Taylor whiskey as E.H. Taylor's straight bourbon. Marion Taylor's Fine Old Kentucky Taylor was a cheaper whiskey and could be sold at prices at which E.H. Taylor could not afford to sell his whiskey, and his deceptive advertisements confused consumers.[201]

In compliance with the courts, Marion Taylor made sure that his Fine Old Kentucky Taylor whiskey was advertised as a blended whiskey and

6 THE WINE AND SPIRIT BULLETIN.

OLD TAYLOR

THE PREMIER KENTUCKY WHISKEY.

The Signature of E. H. TAYLOR, JR. & SONS is always on the Genuine.
INCORPORATED.

E. H. TAYLOR, JR.
Of Taylor Whiskey Fame.

EVERY BOTTLE OF GENUINE TAYLOR BEARS THIS PHOTOGRAPH.

THE COURTS OF
KENTUCKY
DECIDE AS TO
TAYLOR
WHISKEY

OLD TAYLOR IS *The Premier Kentucky Whiskey.*

See that your barrel is *Double Stamped*, and that the *Tax-Paid* brand contains the autograph signature of *E. H. Taylor, Jr. & Sons.*

BEWARE of SINGLE STAMP imitations of OLD TAYLOR.

JUDGMENT

Of the Franklin Circuit Court, April 9, 1891 (Excerpt)

"Said defendants, and each of them, and all their servants, agents and employees are perpetually enjoined and restrained from branding, stamping or in any way marking the packages containing any of their whiskies with the word TAYLOR WHISKEY, and from representing or describing any of their whiskies as TAYLOR WHISKEY, either by brands, signs, labels, show cards or advertisements in newspapers or trade journals or trade reports, or in any way whatever."

OPINION

Of Kentucky Court of Appeals, January Term, 1894, Affirming Franklin Circuit Court, (Excerpt)

"The judgment is also to be approved in denying to the appellants the use of the words TAYLOR or OLD TAYLOR as brands for their whiskies, and in confirming such use to the appellees."

(E. H. TAYLOR, JR. & SONS.)

THERE IS BUT ONE
OLD TAYLOR
DISTILLERY IN KENTUCKY.

E. H. Taylor, Jr. & Sons
DISTILLERS, FRANKFORT, KENTUCKY.
INCORPORATED.

THERE IS BUT ONE
OLD TAYLOR
WHISKEY DISTILLED IN KY.

Edward Taylor's advertisement in *Wine and Spirit Bulletin* from 1903 arguing his side of the lawsuit.

distinguished the whiskey from his Old Charter brand, which was a straight bourbon.[202]

Edmund Taylor nevertheless continued to sue Marion Taylor. The Court of Appeals of the District of Columbia agreed that Wright & Taylor used "Kentucky Taylor" and "Old Taylor" deceptively by advertising "pure whiskey" when it should have labeled its Fine Old Kentucky Taylor as a blend. Under the new Pure Food and Drug Act of 1906, the courts noted that Congress sought to suppress "the manufacture and sale of adulterated foods and drugs and also to prevent their misbranding." The U.S. attorney general required that rectified whiskey be sold as "imitation."[203]

The lawsuits did not hurt Marion Taylor's image in Louisville. In 1905, the citizens of Natchez, Mississippi, sent out an earnest appeal for help. Yellow fever had stricken the city, and Charles Miller, president of the Relief Association of Natchez, wrote to Marion Taylor and the directors of the Louisville Board of Trade, asking for help. Marion Taylor appointed a committee to solicit funds to be sent to Natchez. Wright & Taylor gave $100 immediately to the fund, and Taylor had assurances from Louisville's businessmen that they would make liberal subscriptions to the fund. The board of trade urged Louisville citizens to donate to the fund to help their sister city.[204]

In 1906, Marion Taylor was elected president of the National Wholesale Liquor Dealers Association. He was a member of the board of waterworks. He was a Democrat, and in honor of his services, Louisville Democrats sent him to the Democratic National Convention as a committee man from Kentucky. He was defeated by Johnson N. Camden.[205]

In 1916, he was a member of the Democratic City and County Committee from the Fifty-Fifth Legislative District, consisting of the Sixth and Seventh Wards. He was a member of the Pendennis Club and was elected president in 1915. He was also the first president of the Louisville Country Club. He was the oldest director in point of service on the Louisville Board of Trade. He took an active interest in every movement launched by the organization for the city's welfare. He was the director of the board of trade for twenty years. He officiated as chairman of the entertainment committee for twelve years.[206]

Marion Taylor was one of the largest holders of real estate in the city. His investments were said to run into the millions. After he sold his interests in the whiskey industry, in 1919, he purchased the Marion E. Taylor building, formerly the Paul Jones building, for $1 million, which was the largest realty transaction for a single piece of property in the city's history. He bought the

Marion Taylor monument, Cave Hill Cemetery. *Photo by author.*

Marion Taylor tombstone marker, Cave Hill Cemetery. *Photo by author.*

eight-story office building, which occupied the west side block on 312 South Fourth Street between Liberty and Jefferson Streets. He bought the building from Lawrence Jones and the estate of Saunders Jones.[207] He also bought the Atherton building, located at 610 South Fourth Street, which he named after his wife, Frances.

He married Frances Maize Taylor. He had two half-brothers, Ernest Miller and Dr. Thomas M. Redd, and several nieces and nephews: E. Leland Taylor, Eugene A. Taylor, Watterson Miller, Marion Miller, Kent Miller, Molnery Miller, Mrs. Francis Cole, Mrs. Marion T. Davidson and Mrs. Mildred Taylor.[208]

Marion Taylor died of heart disease at his home on 1378 Third Street on July 1, 1921, at the age of sixty-eight. Funeral services were held at Christ Church Cathedral, and he was buried at Cave Hill Cemetery, Section I, Lot 65, Grave 4. His active pallbearers were Attilla Cox, W.C. Semple, W. Pratt Dale, Owsley Brown, J. Ross Todd and R. Baylor Hickman. His honorary pallbearers were Henry Watterson, Oscar Fenley, Charles Mengel, Charles Grainger, Levi Bloom, John Caperton, John Atherton, William Kaye, Alex Humphrey, Peter Lee Atherton, Jeff Stewart, William Heyburn and George Braden.[209]

Marion Taylor left an estate worth $1.5 million in personal property and $23,000 in real estate, of which half of his estate was left to his wife, Frances Taylor. The remainder of his estate was left to relatives and servants, as well as to a Christmas fund. The fund contained $20,000 and was to be called the Marion E. Taylor Christmas Fund. A special committee was appointed to distribute the money to the poor children of Natchez, Mississippi.[210]

Marion Taylor's will made sure that Mrs. Frances Taylor retained controlling interest in the firm of Wright &Taylor, which owned the Marion E. Taylor building and the Francis building. Prohibition put an end to the Wright & Taylor distillery.

CHAPTER 16

NATHAN URI

Parker Rye

Nathan Uri was born in Paducah, Kentucky, in 1852. His early education took place in the local schools, and when he was old enough, he entered Cincinnati High School. He graduated high school while receiving many honors.[211] His father, Morris Uri, was born in 1819. He received a good education, but in 1848, he left his home in Hechingen, Hohenzollern, Germany, and came to America. He was a peddler in New York and settled in Paducah. He formed a partnership with his brother, Abraham, and opened a country store. They prospered, moved to Louisville and opened a wholesale dry goods business with Israel Heyman under the name of Heyman and Uri, located at Main Street near Fifth.[212]

When the Civil War broke out, all trade with the South was cut off, and the store went bankrupt; Morris moved back to Paducah and reopened his old business and became partners with the Wolff brothers. During the Civil War, Confederate general Nathan Bedford Forrest took all of Uri's stock at his store. The Wolff Brothers firm was dissolved, and Morris moved to Louisville. He entered the wholesale boot and shoe business. The business later went into receivership.[213]

Morris again moved back to Paducah, and with the help of Bamberger, Bloom & Company, he opened a dry goods store. On July 22, 1871, his first wife died, and a year later, Morris died. Nathan Uri, who was twenty-one at the time, took charge of the business.[214]

Nathan's family became associated with Isaac and Bernard Bernheim. In 1874, Uri's sister, Amanda, married Isaac Bernheim. In 1872, Isaac and

Nathan Uri. *From Ben La Bree's* Notable Men of Kentucky at the Beginning of the 20th Century *(1902).*

Bernard Bernheim formed a wholesale liquor sales firm in Paducah, Kentucky. In 1875, Nathan Uri became a partner in the firm, and he was elected vice-president of the company. Isaac Bernheim and Uri realized that Louisville was a larger field for operations and decided to move their operations there. They moved to a six-story building, 32 feet by 210 feet, at 135–37 Main Street. The company carried in stock nearly all of the celebrated Kentucky bourbons and ryes. It was in an excellent position to offer its goods at rock-bottom prices. In 1877, Uri married Addie Levy of Paducah and had three children: Morris, Walter and Mrs. Ella Uri Thalheimer, who was married to Milton Thalheimer.[215]

In 1879, Uri and the Bernheim brothers trademarked their whiskey I.W. Harper and Old Continental. Both of their bourbons were considered to be equal to the finest sour mashes of the times before the Civil War. They boasted that they made their whiskies exactly as Kentucky whiskies were made in the olden times, when two or three gallons were all anyone expected to get from a bushel of grain. They employed twenty-two salesmen, who covered the most of the North, and were expanding into the southern parts of the country.[216]

In 1888, the Bernheims and Uri bought the Pleasure Ridge Park Distillery and renamed the company Bernheim Distillery Company. The distillery could produce ten thousand barrels of sour mash bourbon and rye whiskey every year.

Uri did not remain with the Bernheim brothers for long, and in 1893, he decided to withdraw from the firm and go into business for himself. He continued in the whiskey business after leaving the firm and organized N.M. Uri & Company, with his headquarters on Main Street between Second and Third Streets. He bought the International Distillery at Hunters Station, Kentucky, and his trademark brand was Parker Rye. His own whiskey distillery was known throughout the country by whiskey men and was very popular.[217]

Uri was always prominently identified with every movement for the betterment of the poor. He was a member of the Adath Israel Temple. He belonged to the Standard Club, which was the most exclusive Jewish club

N. M. URI & CO.

Fine Whiskies

LOUISVILLE, KY.

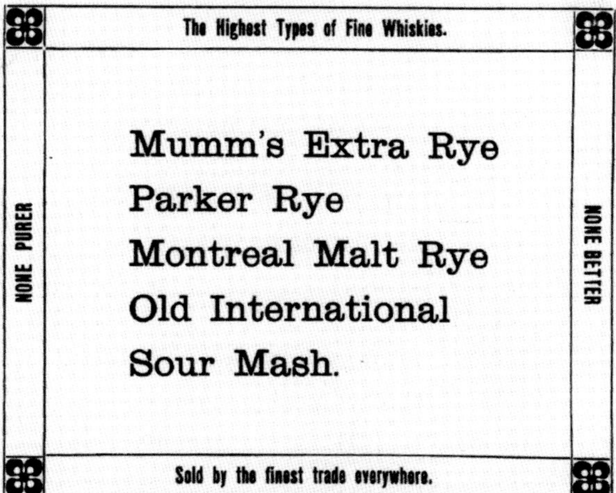

——PROPRIETORS——

International Distillery,

REG. NO. 371.

Hunters, Nelson County, Ky.

ADDRESS ALL COMMUNICATIONS LOUISVILLE, KENTUCKY.

Nathan Uri Company advertisement, 1903. *From* Wine and Spirit Bulletin *(1903).*

in the city, and was identified with the Mystic Shrine and the Louisville Lodge of Elks.[218]

He took a great interest in the theatrical arts. He often stated that nothing was as improving to the mind as a good play of the right kind, and two years before his death, he backed his son Morris Uri in a theatrical venture. He was proud when his son became successful and brought one of his shows to Louisville for three performances.[219]

As a member of the board of aldermen, Uri waged a strong fight for better streetcar service, and conditions in the city improved because of his efforts. He was highly pleased at the numerous improvements instituted by his company and often told his friends that he expected to live long enough to see Louisville have the finest streetcar service of any city in the United States.[220]

He took an active interest in athletics of all kinds and was very fond of baseball. He was a regular attendant at the games when Louisville was represented in the big leagues. He was disgusted when the city was relegated to the rear but stated to friends that any time Louisville had a chance to secure a big-league franchise again, he would back the city.[221]

Uri served on the board of trustees for the Louisville Free Public Library and was a strong advocate of providing branch libraries throughout the city for the poor. He was a friend of the common people. He was also a Democrat and served on the board of aldermen. He was a great reader of books and had one of the finest libraries in the city in his home.[222]

A friend once stated, "Uri was a man of strong convictions, both in politics and business. He always took an intense interest in civic affairs and during the past ten years had been more or less active in public life, serving faithfully and unselfishly with the knowledge that he had done something for the city as his only hope of reward. He led as a member of the Board of Alderman and as one of the finest trustees of the Louisville Free Public Library he set an example for zeal and efficiency which went far toward making the original board of trustees a standard for that organization. Louisville can ill afford to lose men of Uri's character."[223]

On February 22, 1909, Nathan Uri died at his home at 1115 South Fourth Street, on the east side of Fourth Street, between St. Catherine and Oak Streets, from Bright's disease. When death came to Uri, he was surrounded by his immediate family members, except Mrs. Ella Uri Thalheimer, his only daughter, who was in Denver at the time and ill herself; she died a year later on August 3, 1910. Morris Uri, the eldest son, who was engaged in the theatrical business in New York, was in Louisville at the time of his father's death, remaining in Louisville after the concluding performance of *The Three Twins*. Uri's funeral was held in his home, and he was laid to rest at the Adath Israel Cemetery in Louisville. After Nathan's funeral, Walter Uri became president of the company.[224]

On April 2, 1915, at 10:00 p.m., a fire started on the third floor of the Uri building, a four-story structure. The firemen arrived within seconds after the fire alarms were pulled, but they arrived just in time to see the flames from 20,000 to 25,000 gallons of whiskey encircle two buildings. The buildings extended from Main Street to Washington Street. The roar of the explosion could be heard a half-dozen blocks away. An extra fire engine was placed on the Crescent Hill pumping station to supply firemen with more water. Firemen estimated that 1.6 million gallons of water were put on the flames. A third engine was also engaged in putting out the fire. Four alarms were sounded, resulting in sixteen engine companies and three hook and ladder companies responding to the alarms. Walter Uri, the president of the company, estimated the loss on stock at $60,000, which was covered by insurance. The building they owned belonged to Samuel Ouerbacker, and the loss was estimated at $20,000, which was also covered by insurance. P.W. Hardin—president of Hardin, Hamilton & Lewman—estimated his loss at $40,000. The building, which belonged to Stegs heirs, was estimated at $20,000. The seed company had a large stock on hand. W. Ruedeman, president of the H.A. Thierman Company, located at 219 West Main Street, estimated his loss at $1,500 due to water damage. None of the Thierman whiskey was damaged. The roof of the O.K. Stove and Range Company, located at 220–22 West Main Street, caught on fire, but the fire tower extinguished the flames. Walter Uri stated to the local newspaper that there was no interruption of his firm's business; offices were opened in a convenient building, and negotiations for a larger warehouse were already underway.[225]

Walter Uri ran the company until Prohibition closed the doors of N.M. Uri & Company. After the firm closed, Walter was a partner in a mortgage company with offices in the Realty Building. He was also former president of the West Baden Hotel Company at West Baden, Indiana. On January 15, 1929, at the age of thirty-nine, he died from influenza at his home on 521 West Ormsby Street. He was survived by his wife, Sonia, and his brother, Morris. His body was taken to Cincinnati for cremation. He had one daughter, named Alexandra.[226]

Morris Uri's interests were in two areas: baseball and theater. Although most of his New York producing ventures were labeled financial flops, his partner, Joe Gaites, produced a musical comedy hit twenty years before his death titled *The Three Twins*. During World War I, Morris served as manager of an officer's hotel with the American Expeditionary Forces in France. He was given a substantial trust from his father and used the money to compute team averages and follow individual baseball players. His hours of calculations resulted in the book *Handy*. He also published *Perfect Score Book*. As sponsor and umpire, he took an active part in amateur baseball. Morris Uri died at the age of fifty-three on September 25, 1936, in Asheville, North Carolina. He was visiting Asheville hoping to recover his health. His funeral services were held at L.D. Pearson and Son Funeral Home, and his body was cremated.[227]

CHAPTER 17

FRANK, ROBERT, WILLIAM AND ERNEST S. BONNIE

Bonnie Rye, Joel B. Frazier, Nelson Club, Bonnie Bourbon and Bonnie Malt

Frank W. Bonnie was born in Oxford, Butler County, Ohio, in February 1840 and lived in Ohio until he was twenty-five, after which he moved to Kentucky. His brother Robert P. Bonnie was born on May 5, 1855. Frank settled in Richmond, Kentucky, and went into the dry goods business. He moved to Versailles, Kentucky, where he was also involved in dry goods. He was successful in the dry goods business in Cincinnati for several years. Robert joined his brother in Richmond and moved with him to Versailles. In 1870, Frank moved to Louisville. Soon after he arrived in the city, Frank bought the firm of Heath, Smith & Company, wholesale liquor dealers. Later, he formed a partnership with a Mr. Murphy and founded the firm of Bonnie, Murphy & Company. Within a very short time, he became sole owner of the business and changed the name to Bonnie & Company, wholesale liquor dealers and distillers.[228]

In 1871, Robert joined his brother in Louisville and was employed by Bonnie, Murphy & Company as a traveling salesman; he was also in charge of the road sales in the southern states. In 1879, Robert became a member of the firm, and the company changed its name to Bonnie Brothers.[229] William Oregon Bonnie Sr. was born on December 19, 1845, in Oxford, Ohio, where he received his early education. During the Civil War, he served with Company B, the Sixty-Ninth Ohio Union Infantry. The Sixty-Ninth was organized at Hamilton and Camp Chase in Columbus, Ohio. The regiment fought in the Battle of Stone's River, Tennessee; the Tullahoma Campaign; the Battle of Chickamauga; the Siege of Chattanooga; the

Left: William Bonnie. *From Ben La Bree's* Notable Men of Kentucky at the Beginning of the 20th Century *(1902).*

Right: Robert Bonnie. *From Ben La Bree's* Notable Men of Kentucky at the Beginning of the 20th Century *(1902).*

Battle of Missionary Ridge; the Atlanta Campaign; the Battle of Resaca; the Battle of Kennesaw Mountain; the Siege of Atlanta; the Battle of Jonesboro; Sherman's March to the Sea; the Carolinas Campaign; and the Battle of Bentonville. On May 25, 1865, the regiment participated in the Grand Review in Washington, D.C., and the regiment, along with William Bonnie Sr., was mustered out in 1865. After the Civil War, he joined his brothers in Louisville and became a clerk at Bonnie Brothers.[230] Later, Ernest S. Bonnie joined the firm, and the four brothers ran the company. In 1895, Frank retired from the wholesale liquor business and distilling and sold his interest in the company to William, Robert and Ernest S. Bonnie.[231] Bonnie Brothers made Bonnie Rye, Joel B. Frazier, Nelson Club, Bonnie Bourbon and Bonnie Malt. It was in the wholesale liquor business for thirty-five years and in distilling for fifteen years.

In 1899, Ernest Bonnie sold his interest in the company for $70,000 (worth $2,117,768 today), with various notes bearing 6 percent interest per year, including the brands and whiskey, to William and Robert Bonnie, who continued to run the company under the name of Bonnie Brothers. Soon after the sale of his interest in the company, Ernest S. Bonnie went into

BONNIE BROS.,

DISTILLERS,

Registered No. 6, 5th District Ky.

JOEL B. FRAZIER, NELSON CLUB,

BONNIE BOURBON, BONNIE RYE.

Office 139 W. Main St., Louisville, Ky.

Bonnie Brothers advertisement, 1903. *From* Wine and Spirit Bulletin *(1903).*

the liquor business with former employees W.A. Reisert and O.H. Irvine under the firm of E.S. Bonnie & Company. On March 21, 1903, Ernest S. Bonnie changed its name to Bonnie & Company and began to sell whiskey under the brand name of Bonnie Club. William argued that the use of the label and brand Bonnie Club plagiarized certain labels of Bonnie Brothers. After the two brothers wrote to each other, Ernest S. Bonnie abandoned the use of the label. On April 30, 1903, the Bonnie Brothers firm incorporated under the same name, and the corporation took over the assets and brands of the firm. Ernest S. Bonnie came out with a new label titled E.S. Bonnie & Company Rye. The company dropped the "E.S." and began to sell Bonnie & Company Rye, with a label that was very similar to Bonnie Brothers Rye.[232]

After retiring from Bonnie Brothers, Frank became president of the Louisville Bolt and Iron Company, but the company was not successful and went into bankruptcy. In 1866, he married Bell Frazier of Boone County, Kentucky, and they had two sons, Frank and Frazier Bonnie, and two daughters, Madge Bonnie, who was married to Roscoe Frost, and Mrs. Ada Bonnie, who married Charles Curtis Castle. On January 15, 1904, he died from paralysis at his home on 417 West Ormsby Avenue. He left an estate worth $200,000 (worth $5,642,850 today) and left his money to his wife and children.[233]

On January 14, 1904, Robert Bonnie died from Bright's disease at his home on 1336 Third Street. He was survived by his wife and four children: Mattie Sevier Bonnie, Shelby Bonnie, Robert Bonnie and Sevier Bonnie. He was fond of fishing and hunting and was a member of the Louisville Gun

Robert Bonnie monument, Cave Hill Cemetery. *Photo by author.*

Club. He was one of the best field shots and hunters in the state of Kentucky and won many prizes for his marksmanship.[234] He was laid to rest at Cave Hill Cemetery, Section 215, Lot 91, Grave 1.

After the death of Frank and Robert, William became president of the company. In May 1907, E.S. Bonnie died in Halifax, Nova Scotia. For three years, his health had been declining, and he hoped that by traveling to foreign countries, his health might improve. He arrived in Halifax and telegraphed his brother that he hoped to return to Louisville, since his health was improving, but he died in Nova Scotia on May 15, 1907.[235] After his death, E.S. Bonnie's stock in Bonnie & Company passed into other hands.

In 1914, William Bonnie sued his brother's company E.S. Bonnie for trademark infringement. Bonnie Brothers stated that the Bonnie & Company Rye brand and the label on which the brand appears was made in such a way that constituted unfair competition. William stated that there was no doubt that Bonnie & Company Rye infringed on Bonnie Rye. He also could not understand why E.S. Bonnie & Company used the brand Bonnie & Company Rye because the purchasing public thought they were buying Bonnie Brothers Rye. William stated that "no trader can adopt a trademark so resembling that of another that ordinary purchasers, buying with ordinary caution, are likely misled." He stated that Bonnie & Company Rye was a fraud on the public. He also quoted the Federal Food and Drug Act of June 20, 1906, which stated that "all misbranded items applied to drugs, or articles of food, or articles which enter the composition of food, the package or label of which shall bear any statement, design, or device regarding such article, or the ingredients or substances contained therein which shall be false or misleading in any particular and to any food or drug product which is falsely misbranded as to the state, territory, or country, in which it is manufactured or produced."[236] He also stated that if the packaging containing the label listed ingredients or substances that were false or misleading—such as the bourbon in the bottle actually being made up of compounds, imitations or blends—then the label should clearly have the words *compound*, *imitation* or *blend* listed on it. He also included the secretary of agriculture and secretary of commerce and Labor Act from February 16, 1910, which stated that all unmixed distilled spirits from grain, colored and flavored with harmless color and flavor, either by charred barrel process or by the addition of caramel and harmless flavor and not less than 80 proof, are entitled to the name *whiskey* without qualification. Straight whiskey, rectified whiskey, redistilled whiskey and neutral spirits whiskey are like substances and mixtures of whiskies, and under the law, they must be labeled as "blended whiskies."[237]

William stated that E.S. Bonnie's Bonnie Rye was not a straight whiskey, but rather a blended whiskey of two or more straight whiskies of different ages. William stated that Bonnie Rye made by E.S. Bonnie was a fraud on the public and was misbranded.[238]

The courts found that there was no proof to the effect that other manufactured whiskey was sold under the label of Bonnie Rye. There was no misrepresentation as to the place of manufacture or the real nature of the article. The only misrepresentation was the violation of the Food and Drug Act since the brand was a blend of two whiskies made at different times. They did not think that E.S. Bonnie intended to deceive the public or defraud the public. The courts stated that William Bonnie was not entitled to any relief since he did not "come into equity with clean hands." The courts also stated that Bonnie & Company had used the phrase "Bonnie & Company Rye" for ten years, so William Bonnie's right of action was barred by the statute of limitations. The courts stated that the infringement "of a trademark is a continuing injury, and though laches and limitation may defeat an action for damages, neither laches alone nor limitation is

William O. Bonnie monument, Cave Hill Cemetery. *Photo by author.*

available as a defense to an action for relief by way of injunction for such infringement." The judge affirmed the overruling.[239]

In 1874, he married Laura Moss Terrell, and his second wife was Katherine Whedon. He was a member of the Pendennis Club, the Louisville Audubon Country Club and the Louisville Board of Trade. He also served on the board of park commissioners. On March 1, 1923, William Oregon Bonnie Sr. died from pneumonia, which was complicated by a stroke. He was laid to rest at Cave Hill Cemetery, Section 15, Lot 92, Grave 5. He was survived by his wife and two sons, William Oregon Bonnie Jr. and Herbert Bonnie, and two daughters, Mrs. Carl Langenberg and Mrs. Dorothy Bonnie, who was married to Hugh Caperton.[240]

Prohibition ended Bonnie Brothers and Bonnie & Company. After Prohibition, Bonnie Brothers tried to restart its business but sold out to Schenley. By the 1970s, Bonnie Brothers had closed its doors.[241]

CHAPTER 18

MOSES SCHWARTZ

Sweetwood Distillery Company

Moses Schwartz was born on March 5, 1851, in New York and later moved to Nashville. During the 1880s, he moved to Louisville, Kentucky, and by 1888, he had incorporated the Sweetwood Distillery Company, with a capital of $100,000. His office was located at 126 East Main Street in Louisville, and his distillery and warehouses were located at Twenty-Sixth Street and Broadway. His distillery had the capacity for ten thousand barrels annually. The product of his distillery was fire copper whiskey. His business extended to every part of the United States, with shipments coming directly from his distillery. His secretary was Frederick Weiss, and his treasurer was Jacob Schwartz, his brother. Moses also ran a large wholesale liquor business, with his office and saleroom at 126 East Main Street. Moses had been in the wholesale liquor business for fourteen years. He was known throughout the country and was prominently connected to many important local interests. Moses was a member of the board of trade and the Commercial Club.[242] He became director of the German National Bank and the Germania Vault and Trust Company. He secured large sums of money and gave promissory notes to the German National Bank. Financially, he was indebted to the German National Bank for $100,000. There was no way he could pay off the debt.

On March 25, 1891, the Louisville German Deposit Bank was organized, and Moses Schwartz was elected president of the new bank by the board of directors. The new bank, located on the northwest corner of Fourth and

Main Streets, had a capital of $500,000.[243] Louisville financiers took stock in the new bank, and then the German citizens of Louisville began to flock to the bank with their savings. Hundreds of people became depositors at the new bank. Schwartz was spending money and living like a prince. The deposits in the Louisville German Bank allowed him to take a $100,000 of line of credit from the German National Bank to pay for the money he was taking out of the Louisville German Deposit Bank. Basically, Schwartz borrowed $100,000 from the German National Bank to pay the Louisville German Deposit Bank. Schwartz promised to repay the German National Bank by securing an extension of credit. Once he had secured this credit, the German National Bank would be reimbursed at the expense of the Louisville Deposit Bank. In actuality, Schwartz had no intention of paying the German National Bank back, and the German National Bank was the first to declare bankruptcy. Schwartz never paid the money back to the German National Bank. The Louisville German Deposit Bank also failed. Schwartz had taken money from not only the Louisville German Deposit Bank but also the German National Bank.[244]

On July 25, 1893, the deed of the transfer of the Sweetwood Distillery Company was filed in the Jefferson County Clerk's Office due to bankruptcy. The German Safety Vault and Trust Company was named as assignee of the company's bankruptcy. Fred Weiss, the secretary of Sweetwood Distillery, stated that the firm was perfectly solvent but had become tied up with several of the banks that had closed their doors, including the Louisville Deposit Bank and the German National Bank, and had made an assignment for the benefit of those whiskey owners who owned stock in the maturing barrels of whiskey in the Sweetwood Distillery, who wanted to withdraw their stock in the value of the whiskey. Schwartz made an individual transfer. People came to see Schwartz to learn about the assets and liabilities of the Louisville Deposit Bank and the Sweetwood Distillery Company, but no one could find him. Schwartz had fled town. Jacob Schwartz, the distiller, announced that he had suspended payment for a while.[245] The Sweetwood Distillery had 1,400 barrels of whiskey aging in the warehouses, the most valuable part of its assets. Schwartz's private estate consisted of his home on Ormsby Avenue, some whiskey and stocks in various corporations. Both his house and his stock were seized for his creditors.[246] After learning that Schwartz had taken all their money from the Louisville Deposit Bank and the German National Bank, suicides were common throughout the German neighborhoods, since hundreds of people lost everything.

By September 1893, a conspiracy suggested that the German National Bank loaned Schwartz, of the Louisville Deposit Bank, $235,000 on worthless paper. The German National Bank was reimbursed at the expense of the Louisville Deposit Bank, and when the Louisville Deposit Bank failed, the bank was virtually plundered by Schwartz and his relatives. The German National Bank filed suit to recover the money. At one time, Schwartz was a large depositor at the German National Bank. When the Louisville Deposit Bank was organized, Schwartz ceased to be a depositor at the German National Bank, and the bank wanted to take up some of the paper in his bank. The German National Bank collected its outstanding indebtedness from Schwartz. Schwartz used the money from the Louisville Deposit Bank to pay the German National Bank. The German National Bank claimed that Schwartz did not pay back the Louisville Deposit Bank.[247]

Several days later, on September 28, 1893, Augustus Sharpe sued the German National Bank, Schwartz and the Germania Safety and Trust Company. Sharpe stated that he gave Schwartz a $10,000 promissory note that was never paid. The lawsuit also stated that when Schwartz decided to declare bankruptcy, he paid the German National Bank checks for $35,136, $15,000 and $25,212 from the Louisville Deposit Bank. Sharpe stated that Schwartz's payment to the German National Bank was a fraudulent attempt on Schwartz's part to prefer the German National Bank over his other creditors. Sharpe sued to have the assignment to the Germania Trust Company voided and Schwartz's entire estate turned into cash to be paid out to his creditors equally.[248]

In October 1893, the Kentucky Trust Company sued Schwartz, the Germania Trust Company, the Louisville Deposit Bank and the German National Bank. Schwartz gave a promissory note to the Kentucky Trust Company for $7,500 that was never paid. The Kentucky Trust Company also sued Schwartz for $18,535 worth of bonds issued by the Dark Hollow Quarry Company.[249]

Lawsuits continued to be filed in the courts. In April 1894, the stockholders of the Louisville Deposit Bank sued the German National Bank, Adolph Reutlinger, Moses Schwartz, Germania Trust Company (assignee of the Louisville Deposit Bank) and the Louisville Deposit Bank. The lawsuit charged that Adolph Reutlinger, Albert Reutlinger and Moses Schwartz, each acting as officers of the German National Bank and on their own individual behalf, fraudulently conspired and allied together to enter into a fraudulent agreement and conspiracy for the purpose of defrauding and deceiving the stockholders of the Louisville Deposit Bank. Schwartz was a director at the German National Bank and director of the German National Insurance Company, of which Adolph Reutlinger was president. Schwartz was also director of the Germania Trust Company, while Adolph Reutlinger was also president. Schwartz, Adolph and Albert Reutlinger were joint owners of $30,000 of stock and $10,000 in bonds of the Oregon Gold Mining Company. The plaintiffs stated that Schwartz owed $223,000 to the German National Bank, for which the bank was holding worthless paper. Schwartz's collateral—namely the Oregon Gold Mining Company, Southern Oil and Semaphore stock—were completely worthless. By fraud, the Chemical National and the American Exchange National Banks of New York were procured to discount $80,000 worth of paper that Schwartz and Reutlinger knew to be worthless. As part of the conspiracy, Schwartz was made president of the Louisville Deposit Bank, and with his fellow conspirators of the German National Bank, he unloaded a quantity of Schwartz's worthless paper on the newly formed Louisville Deposit Bank. The stockholders were suing for $275,000.[250]

On May 7, 1894, the Jefferson County courts auctioned off stocks and bonds and collateral on notes, to satisfy the judgment of $19,500 rendered to be paid to the Kentucky Trust Company; they also sold Schwartz's twelve bonds from the Dark Hollow Quarry Company.[251]

In June 1894, the German National Bank sued the Louisville Deposit Bank and Moses Schwartz for $50,000 for five notes. The notes were given by the Louisville Deposit Bank to Schwartz as sureties. The defendants stated that Schwartz obtained all the notes issued by the Louisville Deposit Bank by means of fraud, misrepresentation and deceit.[252]

The lawsuits continued into 1895. In a cross petition and counterclaim, the Germania Safety Vault and Trust Company, an assignee of the Louisville Deposit Bank and a creditor for Moses Schwartz, filed a suit against Augustus Sharpe against the German National Bank, Moses Schwartz and the Germania Safety Vault and Trust Company. The cross petitioner stated that its codefendant Moses Schwartz was indebted to the Louisville German Deposit Bank in the sum of $100,000 to the extent that the Germania Trust Company, as assignee of the Louisville German Deposit Company, was a creditor of Moses Schwartz. A counterclaim against the plaintiff and the cross petition against the defendant Moses Schwartz were joined in a petition against the plaintiff. They asked that the assignment to the Germania Trust Company be voided and that a receiver be appointed to settle the estate of Schwartz. They also asked that the German National Bank be compelled to pay into court the various amounts of money alleged to have been paid to Schwartz.[253]

In 1896, the judge ruled in the case of Augustus Sharpe against the German National Bank and Moses Schwartz. The case was dismissed. The courts stated that Sharpe was not affected by the matters complained of in his lawsuit. Schwartz was insolvent, meaning he was unable to pay his debts. The judge stated in his ruling that the German National Bank and the agreement between the two banks "was to compel the Deposit Bank to become a creditor of Schwartz and to lighten the burden of his indebtedness to the German National Bank to the extent of the amounts upon the payment of which it made its assistance conditional." Or, in other words, it simply unloaded $75,000 of bad debts on the Deposit Bank. In these transactions, Schwartz paid nothing, and his estate was not diminished one cent. "The amounts paid to the German National Bank could never in any way have gone to the creditors because they were paid out of the money of the Louisville Deposit Bank upon an agreement that it should be so paid. Whether or not the assignee of the Deposit Bank, or some stockholder or creditor of that institution, could recover for the benefit of the creditors of that bank upon the facts here is not now submitted for decision, but it is sufficiently clear that Sharpe, a general creditor of Schwartz, is not interested in this matter." The courts essentially stated that Sharpe lost because he was a creditor of Schwartz and the case had to do with the German National Bank recovering money from the Louisville German Bank. The lawsuit had nothing to do with the creditors forcing the banks to pay them back. The lawsuit would tie the Louisville German Deposit Bank as a creditor to Schwartz so the courts could liquidate Schwartz's estate to pay the creditors.[254]

In 1897, the drama continued when Barker McKnight, who was also president of the German National Bank and director of the Germania Vault and Trust Company, charged that Reutlinger had mismanaged the German National Bank; that there was double dealing, misrepresentation and false reports of the German National Bank; and that the bank lost $273,000 before 1894. He also claimed that assets of the German National Bank were padded.[255]

As for Schwartz, he had left Louisville, and in 1901, a report came out of New York that he was running the Manhattan Mercantile Company and was in business on Williams Street but could not be located. The New York police believed that Joseph B. Hart, who was arrested for fraud, was Moses Schwartz. A *Louisville Courier-Journal* reporter showed distiller J.B. Wathen a photograph of the man reported to be Moses. Wathen replied, "Oh, no, that's J.B. Hart of New York. The difference between Schwartz and Hart is that one caught me for $45,000, the other for $1,100." Wathen had sold Schwartz warehouse receipts in the amount of $45,000, with stock in the bank as collateral. Wathen told the reporter that "the stock wasn't worth the paper it was written on."[256] Wathen stated that he saw Schwartz several times in New York, hoping to get a settlement out of him, but Schwartz always claimed that he was impoverished, although the style he was living in was contrary to the fact. He had rooms in a swell apartment and was on "the high wave of prosperity."[257]

In 1902, Moses Schwartz was wanted in New York for alleged forgeries regarding the Seventh National Bank amounting to $100,000 and was under arrest in Philadelphia. He was charged with being a fugitive from New York. R.W. Jones, vice-president of the Seventh National Bank, stated that Schwartz had been manager of the Manhattan Mercantile Company and a wholesale liquor dealer, with offices on 35 South Williams Street. The Mercantile Company failed, and a large amount of the company's paper was being carried by the Seventh National Bank. The company's paper was worthless, and some was even forged. When the bank confronted Schwartz, he immediately left New York, and after spending time in Chicago, he moved to Philadelphia and rented an apartment on South Sixteenth Street, where he was arrested. He was taken back to New York for charges.[258]

Interestingly, by 1911, the *Louisville Courier-Journal* was reporting that Moses Schwartz was a wealthy man, according to Isaac Bernheim, who had returned from a trip to Florida. While he was at Palm Beach, Bernheim saw Moses Schwartz and Mrs. Schwartz. Schwartz was surprised to see Bernheim, since they had not seen each other in eighteen years. Schwartz told Bernheim that his two sons, Charles and Morton, were both multimillionaire stockbrokers in New York. According to Schwartz, when he fled Louisville, his two sons were twelve and fourteen. In 1901, his sons secured employment with John W. Gates, "the big Wall Street plunger." They were quick to learn the financial game. Both sons made successful takeovers on Wall Street. They managed to accumulate enough capital to operate on their own. According to Schwartz, his sons were firmly established in New York. Moses Schwartz invested with his sons and retired from business on a big income.[259]

In 1915, in New York, Ella de Peyster Shoemaker married Morton Lewis Schwartz, son of Moses Schwartz. Moses Schwartz died on

December 20, 1923, at the age of seventy-two in Manhattan and was buried at the Woodlawn Cemetery in the Bronx, New York, Sassafras Plot, Section 120, Schwartz Mausoleum. His wife, Jennie, died in 1935. His son Morton died in 1953, and his son Alexander Charles Schwartz died in 1967. He also had two daughters: Amy Ruth Schwartz Oppenheim, who died in 1955, and Corrine Schwartz, who died in 1971.

CHAPTER 19

PHILIP AND ARTHUR STITZEL AND WILLIAM LARUE WELLER

Old Fitzgerald

Philip Stitzel was born in 1846 in Sausenheim, Germany. At the age of eleven, he came to America with his father and two brothers, Frederick and Jacob.[260] Jacob Stitzel was born on December 25, 1847, in Germany, and Frederick Stitzel was born on July 4, 1843, in Rheinland-Pfalz, Germany. The family moved to Louisville shortly after their arrival in America. In 1872, Philip and his brothers engaged in the distilling business. In 1879, Frederick invented and patented the ricking system, which was the system of wooden barrel racks that held his whiskey barrels while they aged in his warehouses. The system allows air to circulate around all the barrels in the aging process and is currently used in every distillery.[261] Frederick was connected with the Newcomb-Buchanan Distillery Company and later formed a partnership with Frederick Adams and others in the Anderson and Nelson Distillery. He sold the property to the Kentucky Distillery and Warehouse Company. Frederick Stitzel later became president of the Stitzel Brothers Distillery. Jacob Stitzel was the manager of the Glencoe Distillery.[262]

In 1902, Philip sold the Stitzel Brothers Company to the Kentucky Distillery and Warehouse Company. His brother Jacob Stitzel later bought the old Stitzel Brothers Distillery at Twenty-Sixth and Broadway.[263]

In 1903, Philip and his son Arthur Philip Stitzel started the Stitzel Distillery Company at 1033 Story Avenue. Arthur was born on March 19, 1875, in Louisville, Kentucky, and was educated in the local schools. Arthur became president of the company, and his father was secretary. In 1904, at the age of fifty-seven, Philip Stitzel died from heart trouble at his home on 2341 West Chestnut Street in Louisville. He was buried at Cave Hill Cemetery, Section 14, Lot 209, Grave 1. He was a member of the Knights of Pythias, the Masons and the Odd Fellows. By the time he died, he had amassed a large estate, which was left to his eight children: Arthur, Philip, Fannie, Alma, Toka, Edna, Ellen and Lillian. He gave four shares of stock in the Stitzel Distillery Company of par value of $1,000 per share to each of his eight children. The rest of the estate, including an interest in the distillery and two houses and lots, went to his wife, Elizabeth. After Philip's death, Arthur continued his father's business until Prohibition.[264]

During Prohibition, Arthur managed to sell medicinal whiskey. Congress had enacted a law permitting consolidated warehouses where bourbon barrels could be stored under government supervision. Stitzel managed to persuade other companies to store their whiskey in his warehouses, so the government issued him a temporary consolidation warehouse permit. His company's main brand was Old Fitzgerald, and the distillery became known as the Old Fitzgerald distillery. Old Fitzgerald was first distilled by John E. Fitzgerald in Frankfort, Kentucky. In 1884, S.C. Herbst of Milwaukee registered Old Fitzgerald and acquired the distillery.[265]

William Larue Weller was born in Larue County on July 26, 1825. He was the son of Samuel and Phebe Larue. He grew up on the family farm, but in 1844, he came to Louisville. When the Mexican-American War broke out, he enlisted with the Louisville Legion. After the war, he returned to Louisville, and on January 9, 1849, he went into the whiskey business with his brother, Charles David Weller. They started W.L. Weller and Brothers Wholesale Merchants in Foreign and Domestic Liquors. His business was located on Main Street between Sixth and Seventh Streets, opposite the Louisville Hotel. Charles was a traveling salesman. On April 25, 1849, he married Sarah B. Pence of Shelby County. They had seven children: George Weller, W.L. Weller, John Weller, R.E. Weller, Mrs. A.R. Sullivan, Mrs. W.R. Blue and Mrs. William Coldewey. In 1854, a typhoid epidemic killed Samuel and most of his family in Larue County. He was the founder of the Louisville Baptist Orphans Home and a leading member of the Walnut Street Baptist Church. During the Civil War, William's brother John and his eldest son, George, joined the Confederacy. John Weller became a captain in Company D, Fourth Kentucky Infantry (CS), and George joined the Fifth Louisiana Infantry. On July 1, 1862, his brother Charles was robbed and murdered while collecting bills in Clarksville, Tennessee.[266]

After the loss of his brother in the firm, William soon had a new partner, James P. Buckner, and the company changed its name to W.L. Weller and

Philip Stitzel monument, Cave Hill Cemetery. *Photo by author.*

Philip Stitzel marker, Cave Hill Cemetery. *Photo by author.*

Buckner. The partnership lasted until his son George came back from the Civil War and joined the company. By 1870, Buckner had left the company, and the company changed its name to W.L. Weller and Sons. On September 1, 1871, George Weller opened W.L. Weller and Sons on Second Street. The firm was a redistilling and rectifying company. By 1872, W.L. Weller was making $1 million per year. George's distillery was one of the most extensive and, without exception, one of the best and most completely equipped on the continent. He introduced into his business every modern invention that had been tested. He had two large stills of double plate copper and two copper columns, stated to be some of the finest stills in the United States. The columns were thirty-six feet high and thirty-six inches wide, with thirteen separate chambers, each one a separate still. Every pipe and still was made of copper. The stills and columns had glass registers and hydrometers. The high wine was poured from the first floor into large wells in the cellar, from which the liquid was passed into the stills. The vaporization of the spirits through the columns removed every objectionable element. After the vaporizing process through the column, the liquor was condensed and received into tubs on the first floor. A steam engine operated the steam-powered elevators and the force pumps. The main force pump had a capacity of twelve and a half barrels of water per minute. Every barrel was put through a steam bleaching process before use. George Weller made bourbon, rye, cognac, French and neutral spirits, gin and brandy.[267]

In 1873, the rectifying house of W.L. Weller and Sons at 24, 26 and 28 Second Street, between Main and the Ohio River, caught on fire. George Weller arrived on the scene and took out the papers and books from the safe. He estimated the loss at $55,000, but he had insurance to cover $37,000. He had twenty-five thousand gallons of spirits in the building with 140 barrels ready to ship.[268] In 1875, the distillery department of W.L. Weller and Sons, on the corner of Main and Brook Streets, burned, with a loss of $20,000, which was partly covered by insurance. In 1877, the W.L. Weller rectifying establishment was damaged by fire.

In 1893, nineteen-year-old Julian Van Winkle started as a whiskey salesman for W.L. Weller Distillery. In 1894, he graduated from Centre College. He was born in Danville, Kentucky, on March 22, 1874. His father, John S. Van Winkle, and his uncle, Ephraim Van Winkle, were both lawyers and Kentucky secretaries of state. As a legislator, Ephraim helped develop Kentucky's public school system.[269] In 1896, William Larue Weller retired and left his business to his sons George and John C. Weller. On March 23, 1899, William died at the Ocala House in Ocala, Florida. He had been

GEO. P. WELLER. ESTABLISHED 1849. JNO. C. WELLER.

W. L. WELLER & SONS,

DISTILLERS

..AND..

Wholesale Liquor Dealers,

LOUISVILLE, KY.

Harvard Club, Mammoth Cave and Silas B. Johnson Whiskies

PRODUCERS OF FINE BLENDS.

W.L. Weller advertisement, 1903. *From* Wine and Spirit Bulletin *(1903)*.

suffering from asthma and hoped that the different climate would help. His brothers Captain John H. Weller and Jacob Weller received the word of his death. He was buried at Cave Hill Cemetery, Section 5, Lot 308, Grave 3.[270]

In 1908, W.L. Weller & Sons was incorporated with a capital stock of $25,000, divided into shares of $100 each and with a maximum debt of $50,000. The incorporators were George P. Weller, George Larue Weller and Alexander Farnsley.[271]

In 1909, eight people were injured while battling a fire that destroyed the W.L. Weller and Sons business on 121 West Main Street. The loss was estimated at between $75,000 and $100,000. Julian Van Winkle, who was the secretary for Weller, stated that the loss to the stock was estimated at $50,000, and the building, which was purchased from George Bernheim ten years before the fire, was estimated at $40,000.[272] During that same year, Van Winkle and Alex Farnsley bought the controlling interest of W.L. Weller & Sons.

On November 20, 1913, Jacob Stitzel, manager of the Glencoe Distillery Company, died at the age of sixty-five at his home on 2404 West Walnut Street. In 1910, he had been forced to retire from the Glencoe Distillery on account of ill health. His son Frank Stitzel took over as manager of

W.L. Weller monument, Cave Hill Cemetery. *Photo by author.*

W.L. Weller marker, Cave Hill Cemetery. *Photo by author.*

the Glencoe Distillery Company.[273] Jacob Stitzel was buried at Cave Hill Cemetery, Section 1, Lot 154, Grave 5.

In 1915, W.L. Weller & Sons, which was known as the oldest whiskey house in Louisville, merged with Trost Brothers of Louisville. The consolidation of the two companies involved assets valued in excess of $250,000. Headquarters for the company remained at Weller's location of 121 West Main Street. George P. Weller was president, Julian Van Winkle was secretary and treasurer and Alex T. Farnsley was vice-president. Trost Brothers, which consisted of William and Isaac Trost, had been a wholesale liquor dealer in Louisville for twenty-two years.[274] During that same year, Van Winkle and Farnsley equally held all but three of the company's 206 shares of stock.

On April 1, 1918, at the age of fifty-four, John Charles Weller died from pneumonia at his home on 1972 Deer Park Avenue. He was also vice-president of the American Rubbing Stone Quarry Company of Floyd Knobs, Indiana. He was survived by his wife, Jennie Weller, and his daughter, Mrs. Charles Cimiotti. He was also survived by his three brothers—George P. Weller, William L. Weller and R. Lee Weller—and a sister: Mrs. W.R. Blue. He was buried at Cave Hill Cemetery, Section 5, Lot 308, Grave 8.[275]

During Prohibition, the Weller Company was one of the few distilleries that could sell bourbon for "medicinal" purposes. In 1933, Van Winkle merged Weller & Sons with A.P. Stitzel Inc. and began to build a new distillery in Louisville. Also in that same year, S.C. Herbst sold the brands and inventory of the Old Judge Distillery on Benson Creek west of Frankfort, which made Old Fitzgerald, to Julian Van Winkle and Weller & Sons.[276]

In 1934, master distiller William H. McGill, who had worked for Early Times and Tom Moore Distilleries, joined Stitzel-Weller. McGill agreed with Van Winkle that making bourbon was an art, not a science.[277] In 1937, the new distillery was completed. The new merged company increased its capitalization from $6,000 to $600,000. Julian "Pappy" Van Winkle became president of the new organization, with Farnsley becoming vice-president and Arthur P. Stitzel becoming secretary.[278]

The new plant was built in two years and occupied a twenty-acre site on Dixie Highway at Shively. The plant produced 1,500 bushels of grain every day. The plant produced 150 barrels and seventy-five thousand gallons of whiskey daily. The metal-clad warehouses had a storage capacity of 12,600 barrels. The bottling plant had the capacity of four thousand to five thousand cases daily.[279]

Original Old Mock label. *Author's collection.*

The new Stitzel-Weller company brands included some of the oldest and best-known names of pre-Prohibition days, as well as brands issued for the medicinal whiskey market during Prohibition. Old Fitzgerald became the flagship brand for Stitzel-Weller Inc. The distillery also made Mammoth Cave, Kentucky Oaks, Cabin Still, Old Elk, Belle of Bourbon, Old Mock, National Club, Belle of Jefferson, Old Stitzel and Billie Burke.[280]

In 1940, Stitzel-Weller came up with Rebel Yell bourbon. Mayor Charles Farnsley convinced his uncle, Alex Farnsley, to create a bourbon brand specifically for southerners. Rebel Yell, which had a citrus quality, used wheat instead of rye as the flavor grain and was marketed exclusively to the southern states. The original label had a Confederate soldier, with sword drawn, riding a galloping horse. In 1984, Rebel Yell was expanded into national and international markets, and the Rebel Yell motif was toned down, with soldier's sword becoming smaller and the Confederate uniform became more muted. In the international markets, the Confederate soldier disappeared completely, with only the Confederate flag on the label. Later, the soldier became the silhouette of a cowboy. By 2014, the Rebel Yell had been changed to RY, and the cowboy had been replaced with a musician.[281]

During World War II, the company survived the grain shortages. In 1942, Julian Van Winkle Jr. quit his job as treasurer of the Stitzel-Weller Distillery to become a tank commander in the Pacific. He was badly injured in the Philippines and awarded the Silver Star medal for valor after fighting in the single most critical battle of the entire campaign against the Japanese at Breakneck Ridge. After the war, Julian Jr. came home a war hero, and as soon as he was discharged, he became vice-president of the company.[282]

On July 3, 1940, George P. Weller died. He was later buried at Cave Hill Cemetery, Section P, Lot 252 N½, Grave 2. On September 18, 1940, Frederick Stitzel passed away at the age of eighty-one at his home on 2855 Frankfort Avenue. He was survived by his wife, Emma Laval Stitzel, and five daughters: Mrs. George Mercke, Mrs. C.L. Isaacs, Mrs. S.S. Lord, Mrs. Frank Wheeler and Mrs. William Rodman. He was also the inventor of a railroad signal semaphore. He built a home on Crescent Hill and was one of the first four residents in the neighborhood.[283]

Frederick Stitzel monument, Cave Hill Cemetery. *Photo by author.*

Julian "Pappy" Van Winkle Sr. produced a "wheated" bourbon, which was made from a grain mix that substituted wheat instead of rye in the blend of corn and malted barley. According to legend, in 1954, Pappy gave his bourbon recipe to William Samuels of Maker's Mark Distillery in Loretto, Kentucky, and Maker's Mark soon became one of the most popular premium bourbons. Julian was a firm believer in making whiskey the old-fashioned way, where he simmered grain for three hours without cooking the grain under pressure. He believed that some whiskey makers ruined their products by trying to speed up the process through chemistry. He had a sign over his distillery that stated, "NO CHEMISTS ARE ALLOWED ON THE PLACE."[284]

On October 30, 1941, Alex Farnsley died, and on April 13, 1947, Arthur Philip Stitzel died at Norton Memorial Hospital. Arthur was buried in Cave Hill Cemetery, Section 26, Lot 113, Grave 1.[285]

During the 1950s and 1960s, Old Fitzgerald's reputation grew with Van Winkle's, and he wrote a series of personal columns that ran advertisements in *TIME*, *Newsweek*, *The New Yorker* and other magazines. In 1964, Pappy Van Winkle retired, and his son Julian Van Winkle Jr. became president of the company.

Arthur Philip Stitzel monument, Cave Hill Cemetery. *Photo by author.*

Alex Farnsley monument, Cave Hill Cemetery. *Photo by author.*

On February 16, 1965, Pappy Van Winkle passed away at the age of ninety at his home on 37 Hill Road. At the time of his death, he was the oldest active distillery executive in the nation. He had been in the distillery business for seventy-one years. He was a member of the Louisville Country Club and a member of the Pendennis Club. He was a board of directors of the Distilled Spirits Institute and the Kentucky Distillers Association. He was also a former director of the Louisville Trust Company. He was survived by his wife, Kate Smith. He had headed the fund drive for construction of the Barnett Owen Memorial Wing at Kosair Crippled Children's Hospital.[286] He was laid to rest at Cave Hill Cemetery, Section 30, Lot 5, Grave 1.

During the 1950s, bourbon was in decline, and vodka and gin rose in popularity. In the 1960s, Julian Van Winkle Jr. fought off several takeover attempts, but other stockholders finally forced him to sell his company. In 1972, Norton Simon Inc., a conglomerate that included Canada Dry and Hunt Foods, bought Stitzel-Weller for $20 million. Norton Simon changed the Stitzel-Weller distillery name to Old Fitzgerald, and the company became part of Somerset Imports. In 1984, Somerset Imports was bought by Distillers Corporation Ltd. (DCL). In 1986, DCL was bought by Guinness PLC, which changed its name to United Distillers. The Old Fitzgerald Distillery became part of United Distillers Manufacturing Inc., but in 1993, the company changed the name back to the Stitzel-Weller Distillery. In 1999, Heaven Hill Distillery bought the Old Fitzgerald brand and began selling the bourbon at the Bernheim Distillery in Louisville.[287] Tom Bulleit currently owns the old Stitzel-Weller Distillery, on 3860 Fitzgerald Road in Shively, and makes Bulleit whiskey.[288]

After selling Stitzel-Weller, Julian Van Winkle Jr. started another brand called Old Rip Van Winkle and negotiated an agreement with the new owners of Stitzel-Weller to buy his whiskey. His new company, J.P. Van Winkle and Son, made hand-painted ceramic decanters containing whiskey. Commemorative decanters were sold as gifts.[289]

In 1974, Julian Van Winkle III graduated from Randolph Macon College in Virginia and worked at a Louisville clothing store for a few years, but in 1977, he joined his father in the decanter business. His father developed prostate cancer and left the firm to his son. In 1981, Julian Jr. passed away from cancer. He was buried in the Van Winkle plot at Cave Hill Cemetery, Section 30, Lot 6, Grave 3. In 1983, Julian Van Winkle III bought the old Commonwealth Distillery on the Salt River in Lawrenceburg, Kentucky, to age and bottle his whiskey.[290]

Left: Julian Van Winkle Sr. monument, Cave Hill Cemetery. *Photo by author.*

Below: Julian Van Winkle Jr. plot, Cave Hill Cemetery. *Photo by author.*

Julian Van Winkle III started a business relationship with Gordon Hue Jr., who was a partner in the Cork N Bottle wine and liquor store in Covington, Kentucky. Hue convinced Julian to put some of his fourteen- and sixteen-year-old whiskey stock into bottles instead of decanters and sell the product under the Van Winkle Family Reserve label. Julian Van Winkle Jr. had made the bourbon at the Stitzel-Weller plant during the 1960s.

The new product would be labeled as a fine after dinner sipping bourbon, not a mixer. Julian and Hue also sold the bottles in French cognac bottles with a cork stopper instead of a screwtop. The new bourbon sold for fifteen dollars per bottle when most bourbons sold for ten dollars.

As a private label for Cork N Bottle, the Van Winkle Reserve was the first aged premium bourbon sold since Old Fitzgerald. Julian III was convinced that premium bourbons would one day become popular, so he set aside his best twelve-year-old whiskey stocks to age longer.

In 1996, Julian III bottled and introduced his twenty-year-old Pappy Van Winkle Family Reserve and then sold the bottle for seventy dollars or more. Later that year, Julian won a platinum medal in the World Spirits Championships. His fifteen-year-old Rip Van Winkle won second place. Many have credited Julian III with helping bourbon regain its heritage and reputation as the American native spirit.[291] Julian III, along with his son Preston, currently runs operations with the Van Winkle brand at Buffalo Trace, also known as the George T. Stagg Distillery, in Frankfort, Kentucky. Buffalo Trace produces Old Rip Van Winkle bourbon and rye whiskey, Pappy Van Winkle, Van Winkle Family Reserve and Van Winkle Special Reserve. Buffalo Trace also produces W.L. Weller Special Reserve, W.L. Weller 12 Year, Old Weller Antique and W.L. Weller C.Y.P.B. Buffalo Trace is currently owned by the Sazerac Company.[292]

CHAPTER 20

JOHN THOMAS STREET BROWN & SONS

Old Prentice

John Thomas Street Brown was born in Munfordville, Kentucky, on June 8, 1829. His parents were Major J.T.S. Brown and Elizabeth Creel Brown, daughter of Colonel Elijah Creel of Culpepper County, Virginia. His grandfather and his grandfather's brothers were among the first to move from Virginia to Kentucky and were noted for their defense of the settlers and in aggressive expeditions against the Indians. J.T.S. Brown's father, J.T.S. Brown Sr., moved to Munfordville. He owned several farms, engaged in business and was postmaster for Munfordville for fifty years.[293]

J.T.S. Brown was given a thorough education and learned the business trade at his father's stores. In 1855, J.T.S. Brown moved to Louisville and entered into a business partnership with Joseph D. Allen, an old school friend. Alfred Allen was the senior of the firm, although he was an inactive partner. The firm was called Allen, Brown & Company and dealt in the wholesale liquor business.[294]

In 1856, Brown married Emily Graham, daughter of Andrew Graham, the famous tobacco dealer. A year later, he took full control of the business. Brown had a grain, produce and whiskey business. Foreseeing the coming Civil War, he called in all his accounts and paid off his debts. Although his losses were heavy, he was able to pay off every claim against his business. From 1860 to 1864, the business prospered, and he had a large quantity of Kentucky whiskies stored in his warehouses. The night after the last of his

J.T.S. Brown is seen here in the middle. *From top left to right*: Davis Brown and Creel Brown. *From bottom left to right*: J.T.S. Brown Jr. and Hewett Brown. *From Ben La Bree's* Notable Men of Kentucky at the Beginning of the 20th Century *(1902)*.

whiskey had been stored in the warehouses, a huge fire, which originated in the government building located next to his warehouses, wiped out his entire fortune.[295]

During the Civil War, John was the postmaster of Louisville, and on September 14, 1861, he enlisted in the Fourth Kentucky Mounted Infantry Company C; he was promoted to captain on January 8, 1864. During the Civil War, he sold liquor to the troops. In 1870, he joined his brother George Garvin Brown's business and formed J.T.S. Brown & Brothers. George and J.T.S. called their flagship bourbon brand Old Forester. The firm changed its name to Brown, Chambers & Company and then Chamber & Brown. In 1874, J.T.S. Brown withdrew from the firm to start his own bourbon company.[296]

After Brown established his own wholesale liquor business, he allowed his sons to join the company. His son Graham Brown was the first to work for his father's business. He was followed by Davis Brown and J.T.S. Brown Jr. Davis joined his father's company after graduating from Male High School. The name of the company was changed to J.T.S. Brown and Sons. Graham left the firm to enter college, and his two youngest brothers Creel and Hewitt Brown took over the company. They opened a second location on Whiskey Row that served as their warehouse and bottling facility.

In 1894, the Brown brothers decided to buy their own distillery. They bought the Waterfill Distillery (McBrayer) in Lawrenceburg, Kentucky, which is currently the Wild Turkey Distillery. The distillery made 150 bushels per day, but the Browns increased the output to 300 bushels per day.[297] Their flagship brand was Old Prentice, named after the chief editor of the *Louisville Courier-Journal* newspaper, George D. Prentice. Old Prentice was their best-selling brand, but they also made J.T.S. Brown, Old Lebanon Club and Vine Spring Malt. The Old Prentice brand had the iconic red ship's bell.

In 1904, Emily Brown passed away, and J.T.S. Brown died in 1905. He was laid to rest at Cave Hill Cemetery, Section 1, Lot 84, Grave 2. At the time of his death, Brown possessed a large amount of Louisville real estate. By the time of his death, he had already shifted the greater part of his business to his five sons: Davis, J.T.S. Jr., Creel, Graham and Hewett. Although J.T.S. Brown was not an active member of the business at the time, he had been serving as a counselor and adviser. He also had two daughters: Carrie Irwin and Emily Brown.

When he died, J.T.S. Brown's estate was worth $500,000. He left his estate to his seven children. He also left $3,000 to Cave Hill Investment

Top: Original printers plate for Old Prentice Distillery. *Author's collection.*

Bottom: J.T.S. Brown monument, Cave Hill Cemetery. *Photo by author.*

Company, the interest from which was set aside to care for the family burying ground at the cemetery.[298]

In 1907, J.T.S. Brown & Sons was incorporated with a capital of $500,000, which was divided into shares of $100 each. Davis had 1,665 shares, Creel had 1,785 shares, J.T.S. Jr. had 1,885 shares and Hewett had 1,285 shares.[299] In 1910, the Brown brothers built another distillery in Lawrenceburg, Anderson County, Kentucky, on the north side of Bond Mills Road called the Old Prentice Distillery, which was constructed in the Spanish Mission style.[300]

During Prohibition, both plants in Lawrenceburg were closed. On June 15, 1932, Davis Brown died at the age of fifty at his Lawrenceburg, Kentucky summer home, Montrose. His funeral services were held at his home, and he was buried at Cave Hill Cemetery, Section 1, Lot 85, Grave 6. He was also vice-president of the Louisville Water Company. When he died, he left his wife, Agnes Fidler Saddler Brown, with a management stake in the business, making her the first woman in American history to have a role in running a distillery.[301] During Prohibition, J.T.S. Brown & Sons gained a permit to sell medicinal bourbon from its warehouses.

On September 26, 1933, Creel Brown of Anchorage announced that he was going to start a new company called J.T.S. Brown's Sons Distillery Company, capitalized at $500,000. He planned to buy the old Early Times Distillery property in Bardstown, Kentucky, for $40,000. He was to become president of the company, with his son, Creel Brown Jr., as vice-president and Charles Farnsley as a member of the board of directors.[302] On March 28, 1935, Creel Brown Sr. died at Kentucky Baptist East Hospital after suffering from appendicitis. His funeral services were held at his home on Osage and Stone Gate Road in Anchorage, Kentucky. He was buried at Cave Hill Cemetery, Section 1, Lot 85, Grave 8.

After Prohibition, the Spanish Mission–style distillery plant was refurbished and started production in 1933 as the Old Prentice Distillery. In 1941, the plant was purchased by National Distillers Products Corporation, and in 1944, the plant was sold to Charles Grosscurth. In 1946, the plant was sold to Calvert Distilling Company and began producing Four Roses Bourbon.[303]

After Prohibition, the Gould brothers bought the McBrayer distillery and renamed it the J.T.S. Brown Sons Distillery until 1971, when the Austin Nichols & Company bought the distillery and renamed it Wild Turkey Distillery.[304]

On December 6, 1940, at the age of seventy-nine, John Thompson Brown Jr. died at his home in Daytona Beach, Florida. He had moved to Daytona Beach in 1926. He was the vice-president of the Louisville Taxicab and Transfer Company, also known as Yellow Cab. He was buried at Cave Hill Cemetery, Section 11, Lot 101 NE½, Grave 2.[305] On September 9, 1946, Graham Brown died at the age of eighty-eight in Shelbyville, Kentucky. He was vice-president of J.T.S. Brown & Sons. He was buried at Cave Hill Cemetery, Section 1, Lot 84, Grave 10.[306] On November 7, 1952, Hewett Brown died at the Doctors Hospital in Miami, Florida. He was eighty years old. He had moved to Florida and was formerly president of the Coral Gables First National Bank and chairman of the Pan-American Pictures Corporation. He was educated at Male High School and later took courses in banking, commercial law and financing. He was married to Laura Phillips Avritt of Louisville. He was also a member of the Louisville Board of Trade, the Elks Club and the Pendennis Club. He was buried at Cave Hill Cemetery, Section 26, Lot 14, Grave 1.[307]

In 1955, Creel Brown Jr. sold the business. The Heaven Hill Distillery continues to make J.T.S. Brown & Sons bourbon.

CHAPTER 21

GEORGE G. BROWN AND SONS

Brown-Forman Corporation

George Garvin Brown was born in Munfordville, Kentucky, on September 2, 1846. His parents were J.T.S. Brown Sr. and Mary Garvin Brown. He gained his early education in the schools of his native town, but when the Civil War broke out, his education came to a halt because so many young men had joined to serve either in the Confederacy or the Union. In 1863, Brown moved to Louisville, where he attended high school for eighteen months. After graduation, he returned home. In September 1865, he returned to Louisville, where he found employment as a clerk in a wholesale drug firm named Arthur Peters. He also engaged in the tobacco business on Main Street. Three local banks failed in Louisville in one day, and the tobacco firm Brown was involved with went bankrupt, leaving Brown with $18,000 in debt. Although the bankruptcy court wiped out all debts, George kept a record of his creditors. From an insurance policy, he was able to pay off all his debts from the bankrupt company. As a token of appreciation, the Citizens National Bank of Louisville, the principal creditor of the firm, presented him with a loving cup.[308]

Later, he engaged in the whiskey brokerage business and formed a partnership with Henry Chambers. In 1870, George G. Brown borrowed and saved $5,500, and with his brother, John Thomas S. Brown, they established the firm of J.T.S. Brown & Brothers. Henry Chambers was made a partner in the firm.[309]

George G. Brown changed the whiskey business by putting bourbon in bottles. Before the concept of bottling bourbon, whiskey was sent in barrels to the taverns or retailers, who would fill a bottle or a jug's worth for a consumer. Along the way to the retailer or tavern, deceptive retailers would dilute the whiskey with water or other substances to increase their profits. George decided that the only way to ensure that his product was of the highest quality was to sell the sealed product in bottles. George and J.T.S. Brown called their flagship brand Old Forester, named after Dr. William Forrester, who was a renowned Civil War doctor in the Union Fifth Kentucky Cavalry. Forrester endorsed the bourbon. Doctors claimed that a barrel of whiskey could be altered, and liquor poured into a jug could also be altered. Since liquor was a major sedative during the Civil War, doctors wanted a reliable source of bourbon that was not altered in any way. Brown put his bourbon in a sealed bottle and added a handwritten label of assurance.[310]

The firm changed its name to Brown, Chambers & Company and then Chambers & Brown. In 1874, J.T.S. Brown withdrew from the firm to start his own bourbon company, and in 1881, Chambers withdrew from the company. Brown added George Forman and James Thompson to the company, and the firm changed its name to Brown, Thompson & Company. In 1872, George Forman joined the firm.[311] George Forman was born on August 7, 1844. His father was Thomas Seabrooke Forman, who was a large baggage and rope manufacturer. His father had moved to Louisville from Mason County in the early 1800s. His brother, James Forman, was a colonel in the Fifteenth Kentucky Union Volunteer Infantry and was killed at the Battle Stone's River, Tennessee, in January 1863. George Forman's wife was Hanna Bartley.[312]

Thompson retired in 1890, and George Brown and George Forman continued the business under the name of Brown, Forman and Company. Their office and salesroom occupied a six-story building at 123 and 125 West Main Street, where they carried a large stock of fine whiskies of which they were sole owners, including Old Forester, Old Forman, Diamond Bluff, Webwood, Beech Fork, Fox Mountain, La Rue, Sidroc, La Clede, Mason Rye, Kentucky Rye, Major Paul, Golden Age, Hillside and O.S.K. They also handled leading brands such as Mellwood, Atherton, Brownfield, Searey, Marion, G.W.S., Sugar Valley, N. Harris and Golden Wedding Rye; in addition, they handled peach, apple and California brandies, sherry, port and Catawba wine. They also carried a large stock of old whiskies in their Louisville warehouses.[313]

In 1894, George Forman became the first president of the National Wholesale Liquor Dealers Association.[314] In 1901, he died and was laid to rest at Cave Hill Cemetery, Section 3, Lot 27, EP Grave 1. He was the

William Forrester tombstone, Cave Hill Cemetery. *Photo by author.*

last of his family. After Forman's death, George Brown purchased the entire business and changed the name to Brown-Forman Company. In January 1902, the business was incorporated, and the distillery operated at St. Mary's, Kentucky. In 1904, Owsley Brown joined the company. Owsley Brown was born in 1879 and was educated at Centre College, and in 1902, he received his law degree from the University of Virginia. He was admitted to the Kentucky bar and practiced as a lawyer for two years before joining his father's company.[315]

The Brown-Forman offices and warehouses were in Louisville. George Brown became president of the company, and Owsley Brown became vice-president, with Vernon Brown as a member of the board of directors. In 1906, George Brown, along with Thomas Gilmore, formulated the plans for the Model License League. In 1910, George Brown wrote a book titled *The Holy Bible Repudiates Prohibition*. It contained all the verses in the Bible where the words *wine* or *strong drink* appeared and tried to prove that the scriptures commended and commanded the temperate use of alcoholic beverages. He believed there was no more moral turpitude in manufacturing and selling any intoxicating liquor than there was in manufacturing and selling any

George Forman tombstone, Cave Hill Cemetery. *Photo by author.*

other product. He realized that man was responsible to God for his every act and that the conditions surrounding every individual act must be taken into consideration in determining whether it is right or wrong. He did not support Prohibition. He believed that when prohibitionist leaders claimed that the Bible sanctioned their movement and claimed the churches as allies, they were guilty of a great wrong.[316]

By the time Prohibition took effect, Brown-Forman had changed its company. It was no longer a rectifier, which blended whiskies from other companies to distill its own bourbon. In 1917, George G. Brown died at the age of seventy. He was buried at Cave Hill Cemetery, Section A, Lot 398, Grave 12. He was survived by his wife, Amelia Owsley Brown; his two sons, Owsley Brown, vice-president of the Brown-Forman Corporation, and Robinson Swearingen Brown; and his two daughters, Mrs. Elizabeth Bodley Hammond, married to Howard Hammond, and Amelia Brown. He also had a stepson, Embry Swearingen, who was president of the First National Bank, and a grandson, Robert Pusey Hasting. After his death, his son Owsley Brown became president of the company.[317] From 1919 to 1933, Owsley Brown obtained one of six permits allowing Brown-Forman to produce whiskey to be distilled for medicinal purposes. In 1923, Owsley Brown bought the Early Times company during Prohibition.[318]

In 1933, with the repeal of Prohibition, Owsley Brown and several other prominent men in the bourbon industry formed the Distilled Spirits Council of the United States, whose mission was to educate the public on the responsible use of alcohol.[319] During that same year, his son William Lee Lyons Brown joined the firm as secretary and director

George G. Brown monument, Cave Hill Cemetery. *Photo by author.*

of the company. William Lee Lyons Brown was born on July 26, 1906. He attended the U.S. Naval Academy and the University of Virginia. He worked for W.L. Lyons & Company, which became J.J.B. Hilliard & W.L. Lyons & Company, a firm that was founded by his great-grandfather, before joining Brown-Forman.[320]

In 1940, the Brown-Forman Company bought the Labrot & Graham Distillery near Versailles that later became Woodford Reserve Distillery. The original distillery was started in 1812 and operated by Elijah Pepper. In 1941, Brown-Forman converted the Old Forester plant to make distilled alcohol for World War II.[321] Also in 1941, W.L. Lyons Brown became vice-president of the company. In 1945, Owsley Brown became president of the company.

In 1951, W.L. Lyons Brown became chairman of the board after Owsley Brown retired. In 1952, Owsley Brown died at the age of seventy-three at Norton's Memorial Infirmary after five years of failing health. He was the great-grandson of Governor William Owsley. His Harrods Creek home was called Ashbourne, where he bred racehorses. He was former chairman of the Distillers Code Authority. He was director of the First National Bank, the Kentucky Trust Company and the American Arbitration Company. He was also president of the River Valley Club and Louisville Country Club. During World War I, he was chairman of the Kentucky Council for National Defense. During World War II, he was a member of the Advisory Committee for Alcohol and Heavy Chemicals. He was president of the Pendennis Club and a member of the Elks Club, the Masonic Order and the Filson Club. He was married in 1905 to Laura Lee Lyons, who died in July 1933. He had two sons, W.L. Lyons and George Gavin Brown II, and a daughter, Amelia Brown Frazier.[322] Owsley Brown was buried at Cave Hill Cemetery, Section 28, Lot 190, Grave 2.

When Owsley Brown died, his son George Gavin Brown II took W.L. Lyons Brown's position as president of the company. In 1955, they built the Early Times Distillery, which later became the Old Forester Distillery, in Shively, Louisville, Kentucky. In 1956, the company bought Jack Daniel's Tennessee Whiskey in Lynchburg, Tennessee.[323]

In 1965, Brown-Forman purchased Korbel California champagnes and brandies. The following year, in 1966, W.L. Lyons retired as chairman of the board, and George Garvin Brown II became chairman of the board. Daniel L. Street became president of the company. Brown-Forman acquired Old Bushmills Irish Whiskey and Pepe Lopez Tequila. In 1968, the company moved its headquarters to 850 Dixie Highway, Louisville, Kentucky. Fontana-Hollywood became a part of Brown-

Owsley Brown monument, Cave Hill Cemetery. *Photo by author.*

Forman. Fontana-Hollywood imported Bolla Italian wines. In 1969, George Garvin Brown II died, and W.L. Lyons became chairman of the board. William Lucas became president and CEO.[324]

In 1971, Brown-Forman acquired Canadian Mist, located in Collingwood, Ontario, and the import rights to the French vermouths of Noilly Pratt. Robinson Brown Jr. was elected chairman of the board after W.L. Lyons Brown retired. On January 5, 1973, W.L. Lyons Brown died. During his lifetime, he was director of the Pennzoil United Inc. and Atapaz Petroleum Inc. and was president of the Ashbourne Realty and Land Development Corporation. He was also former director of the Chicago, Indianapolis and Louisville (Monon) Railroad and the Jefferson Island Salt Company. He was nationally known as a breeder of Shorthorn cattle and was former president of the Shorthorn Breeders Association. He also raced Thoroughbred horses. He was involved in organizational work, was vice-president of the Navy League and a member of the Sons of Colonial Wars, Newcomen Society, River Valley Club, Wynne Stay Club, Louisville Country Club and the Pendennis Club. He was also an endowment member of the Filson Club. He was a member of the Saddle and Sirloin Club of Chicago, the Gulf

Stream Club, Gulf Stream Bath and Tennis Club, the Everglades Club and the Sail Fish Club of Florida in Palm Springs. He had three sons: Martin S. Brown, W.L. Lyons Brown Jr. and Owsley Brown II. He had a daughter, Ina Brown Musselman, and twelve grandchildren.[325] He was laid to rest at Cave Hill Cemetery, Section 28, Lot 190, Grave 4.

In 1971, Brown-Forman sold its interest in Labrot & Graham Distillery. The company eventually repurchased the company. In 1975, W.L. Lyons Brown Jr. became president and CEO. Brown Jr. was born on August 22, 1936. He attended private schools and graduated from the University of Virginia and the American Graduate School of International Management.[326] In 1960, he joined Brown-Forman Corporation. In 1979, Brown-Forman purchased Southern Comfort.

On June 1, 1980, Robinson Brown Jr. retired as chairman of the board. In 1983, Brown-Forman purchased Lenox Inc. Lenox made china, crystal and giftware. The Brown-Forman Company also acquired the Hartmann Luggage Company. W.L. Lyons Jr. was elected chairman of the board. Owsley Brown II was elected president. Owsley Brown II was born on September 10, 1942, and joined the company in 1969. He was a graduate of Yale University and Stanford University. He served for two years in U.S. Army Intelligence from 1968 to 1969. He also served as an officer for the Actor's Theater, the Filson Historical Society and the Louisville Orchestra and served on the Government Commission for the Kentucky Center for the Arts.[327] In 1987, Brown-Forman acquired the distribution rights and trademark for Fontana Candida Italian wines. In 1989, Brown-Forman purchased Crouch & Fitzgerald, the oldest luggage company in the United States.[328]

In 1990, Brown-Forman gained exclusive rights to distribute Brolio wines, from the oldest winery in the world under continuous family control in the United States, and became its marketing agent. Lenox acquired Kirk Stieff Company, the oldest manufacturer of silver and pewter products in the United States. In 1991, Brown-Forman created a wine division and acquired Jekel Vineyards. The wine division of Brown-Forman acquired exclusive rights to Glenmorangie Single Highland Malt Scotch. In 1992, Brown-Forman acquired Fetzer Vineyards of Mendocino, California. In 1993, Owsley Brown II became the CEO, replacing W.L. Lyons Jr.[329] In the same year, Brown-Forman acquired Carmen Vineyards Chilean Wines. In 1994, the company changed its name from Brown-Forman Beverage Company to Brown-Forman Beverages Worldwide. William Street was elected president and CEO.[330]

In 1995, Brown-Forman expanded to India and formed the Jagatjit Brown–Forman India Private, Ltd., with its headquarters in New Delhi, which supplied its product to India markets. It also established the Brown-Forman Wines International Company. In 1996, Brown-Forman restored Labrot & Graham Distillery in Woodford County, Kentucky, located on the first commercial distillery in the state, and reopened the distillery under the name Woodford Reserve. In 1999, Brown-Forman added Sonoma-Cutrer Vineyards and Mariah Wines of California, as well as two Australian wines, McPherson and Owen's Estates.[331]

In 2000, Owsley Brown Frazier, great-grandson of George G. Garvin Brown, retired. In 2001, Brown-Forman purchased Appleton Estate Jamaican Rum. In 2004, Brown-Forman acquired 20 percent of the stock in Finlandia Vodka Worldwide Ltd. Brown-Forman and Altia jointly own Finlandia Vodka. In 2005, Brown-Forman sold Lenox to Department 56.[332]

In 2006, Brown-Forman acquired Chambord Liquor. Also in the same year, the board of directors elected Sandra Frazier, Martin S. Brown Jr. and George G. Brown IV to become directors of the company. All three of the new directors are descendants of George G. Brown. In 2007, Brown-Forman acquired full ownership of Don Eduardo Tequila. In 2007, Paul Varga became chairman of the board to replace Owsley Brown II, and George Garvin Brown IV was elected president of the board of directors.[333]

In 2011, Brown-Forman acquired Maximus Vodka Brand. In the same year, Owsley Brown II died. In 2012, Owsley Brown Frazier died at the age of seventy-seven. In 2015, Brown-Forman elected Stuart R. Brown and Augusta Brown Holland, fifth-generation family members, to the board of directors. In 2016, Brown-Forman sold the Southern Comfort and Tuaca trademarks to Sazerac. In 2016, the board of directors elected Campbell P. Brown, Marshall Farrer and Laura Frazier, fifth-generation descendants of George G. Brown, to the board.[334]

In 2018, Old Forester Distilling Company opened on Whiskey Row in Louisville, Kentucky. Lawson Whiting became CEO, replacing Paul Varga.

CHAPTER 22

THE JIM BEAM FAMILY BOURBON LEGACY

Jacob Beam was the son of a German immigrant and settled near Hardin Creek, Washington County, Kentucky. Jacob built a gristmill and a small kettle-type still on his farm. He grew his own corn on his farm, and in 1795, he began selling his corn whiskey. In 1820, Jacob Beam handed the distillery over to his son David, who expanded the business and began using column stills. In 1853, David's son, David M. Beam, moved to Nelson County, Kentucky, and when the Louisville and Nashville Railroad was completed, he built a line through Nelson County and constructed a larger distillery seven miles west of Nelson County near the railroad. David M. Beam also branded his bourbon, calling the product Old Tub.[335]

David M. Beam's son Roy Beam Sr. was a master distiller and superintendent for nineteen years with the Frankfort Distillers Corporation in Louisville, Kentucky. He was also a master distiller and superintendent for the 21 Brands Distillery Inc., a position he held for three years at the Park & Tilford Distillery Corporation. He lived at 3 Meadow Brook Road in Frankfort, Kentucky, but was a Louisville resident for many years. He died of cancer at Saints Mary and Elizabeth Hospital in Louisville at the age of sixty on June 25, 1959. He was buried at St. Joseph Cemetery in Bardstown, Kentucky.[336]

At the age of twenty-one, John "Jake" Henry Beam broke away from the Beam family distillery in Mooresville and started his own distillery. In 1856, he built the distillery plant near the Early Times station in Nelson County. Jake Beam, brother of David M. Beam and uncle of James B. Beam, called

Above: David M. Beam tombstone, St. Joseph's Cemetery, Bardstown, Kentucky. *Photo by author.*

Opposite, top: Early Times advertisement, 1903. *From* Wine and Spirit Bulletin *(1903).*

Opposite, bottom: "Jack" Beam monument, St. Joseph's Cemetery, Bardstown, Kentucky. *Photo by author.*

his new distillery Early Times Distillery, and the company was incorporated in 1878. Jake Beam's office was located at 100–102 Main Street in Louisville, Kentucky. The distillery had the capacity for distilling 4,500 barrels of whiskey annually. The firm of George Pearce, B.H. Hurt & Company was the selling agent for the sale of Early Times.[337]

During the financial Panic of 1893, which lasted until 1897, Pearce and Hurt gained control of the distillery; B.H. Hurt became president of the company, and Jake Beam became vice-president and distiller. Jack and his son, Edward D. Beam, ran Early Times until 1915, when Jack and Ed died just one month apart from each other. Jack was seventy-six years old when he died, and Edward was forty-two. Both father and son were laid to rest in Bardstown, Kentucky, at the Bardstown City Cemetery. Beam's nephew, John W. Shaunty, became president of the company. The main office was moved from Louisville to Paducah.[338]

During Prohibition, the plant closed, but whiskey was still aging in the warehouses. Brown-Forman bought the Early Times bourbon barrels for "medicinal purposes." Early Times moved from its station near Bardstown to Brown-Forman's warehouses in Louisville. After Prohibition, Brown-Forman resumed making Early Times at its distillery that made Old Forester. As for the Early Times Distillery at the station in Nelson County, J.T.S. Brown bought the property and built a forty-bushel plant known as J.T.S. Brown and Sons Company to produce Old J.T.S. Brown. In 1956, the Browns sold the distillery to Schenley.[339]

Another son of David M. Beam was James B. Beam. James began to work in the distillery at the age of sixteen. When he was thirty years old, James B. Beam became a partner with his brother-in-law, Albert Hart, in the firm of

Jake Beam tombstone, St. Joseph's Cemetery, Bardstown, Kentucky. *Photo by author.*

Beam & Hart in Nelson County, and in 1898, James and Albert formed the Clear Springs Distillery Company near Bardstown, Kentucky. James also built a new rack house to store more of his bourbon.[340]

Prohibition led to the distillery's closure. In 1913, James purchased the Murphy Barber Distillery at Bardstown, and in 1935, James; his son, Thomas Jeremiah Beam; his brother, Park Beam; and his nephew, Carl Beam, rebuilt the Murphy Barber Distillery and organized the James B. Beam Distillery Company. The plant began producing after 120 days. James B. Beam became president of the company.[341]

James B. Beam's brand names Jim Beam and Old Tub became famous. He also served as an official for the Kentucky Distillers Association. He was made an official Kentucky Colonel by former Kentucky governor Ruby Laffoon. In 1947, James B. Beam died at the age of eighty-nine at his home in Bardstown, Kentucky. He was survived by his wife, Mary Montgomery Beam; his son, Thomas Jeremiah Beam; and two daughters, Mildred Beam and Mrs. Margaret Thompson, who was married to Frederick Booker Noe. He also had two brothers, George and Park, and two sisters, Nannie Hart and Sue Wilson. His funeral was held at his residence.[342]

Before he died, James left operations of the plant to his son, Thomas Jeremiah Beam. Thomas expanded the business. In 1950, Booker Noe became a master distiller for the company. In 1954, Thomas opened a second distillery near Boston, Kentucky. During the 1950s, 1960s and 1970s, Thomas's sons, Baker and David, grew the business. In 1945, Jim Beam was bought by Harry Blum, and in 1968, American Tobacco, which later became Fortune Brands, bought the company. On May 2, 1977, Thomas

Beam family plot, St. Joseph's Cemetery, Bardstown, Kentucky. *Photo by author.*

James B. Beam, St. Joseph's Cemetery, Bardstown, Kentucky. *Photo by author.*

Thomas Beam monument, Cave Hill Cemetery. *Photo by author.*

Beam died at the age of seventy-seven at his home on 11 Totem Road. He was a former president of the Kentucky Distillers Association. He was also a member of the board of advisors of the Kentucky School for the Blind and a former member of the executive committee of the University of Kentucky Alumni Association. He was a member of the American Legion Post, the Knights of Columbus, the Pendennis Club, Louisville Country Club and the Society of Colonial Wars. He was survived by his wife, Lucy Kavanaugh, and two sisters: Mrs. Mildred Beam, who married Thomas Spalding, and Mrs. Margaret Thompson Beam, who was married to Booker Noe. His body was laid out at his residence, and his funeral took place at St. Leonard Catholic Church on Zorn Avenue. He was laid to rest at Cave Hill Cemetery, Section 33, Lot 59, Grave 1.[343]

After Thomas Beam's death, Fred Noe II, also known as "Booker," took control of the company. He was the sixth generation of his family to distill bourbon. In 1987, Jim Beam bought National Brands, along with the trademarked brands Old Crow, Bourbon De Luxe, Old Taylor, Old Grand Dad and Sunny Brook. Old Taylor was later sold to Sazerac. In 1988, Fred Noe introduced a small-batch ultra-premium bourbon that was bottled straight from the barrel without filtering, the first of the company's "small batch bourbon collection." In 1992, Fred Noe retired from the distillery, but he continued to act as a spokesman for the company and oversaw the distilling of Booker's Bourbon with the help of his son Fred Noe III. Noe also served as a spokesman for Jim Beam brands worldwide. On February 24, 2004, at the age of seventy-four, Fred Noe II passed away from diabetes-related ailments at his home in Bardstown, Kentucky.[344] His son Fred Noe III is currently the master distiller and global ambassador for Jim Beam. Jim Beam recently opened a Jim Beam Urban Still House on Fourth Street in Louisville, Kentucky, but due to the Covid pandemic, the Still House was closed. The company also built a state-of-the-art visitor's center at its Clermont plant. Although many in the family continue to be master distillers and ambassadors for the Jim Beam distillery, in 2014, the company was sold to Beam Suntory, which is a subsidiary of Suntory Holdings of Osaka, Japan. Currently, Jim Beam is one of the best-selling brands in the world. Thirty descendants of Jacob Beam have been distillers.

CHAPTER 23

DIETRICH MESCHENDORF

Old Kentucky Distillery

Dietrich Meschendorf was born in Germany in 1858. His father was engaged in the pork packing business in a small town and died when his son was very young. After leaving the common schools in Germany, Meschendorf studied in high school and was in the middle of his term when his mother decided to immigrate to America. He was fifteen years old when he arrived in America. Upon their arrival, Meschendorf's mother and her two sons moved to Louisville, and young Dietrich continued his studies in high school; after three years he decided to start his business career.[345]

Dietrich opened a small grocery, and his business was fairly successful; from the profits, he managed to save his money and decided to leave the grocery business and invest in a distillery. In 1893, Meschendorf and his business partner, Charles Lemon, along with others, bought the Old Times Distillery Company. He was made secretary and treasurer. John Roach had established Old Times Distillery in 1869 and sold the distillery to Anderson Biggs in 1878, and when Biggs died in 1889, his widow sold the distillery in 1889 to Meschendorf, Charles Lemon and A.W. Bierbaum and his associates. When A.W. Bierbaum, as president of the distillery, died in 1890, Meschendorf became president of the company.[346]

In 1895, Charles Lemon and D. Meschendorf were sued for trademark infringement. Judge Sterling Toney, in the law and equity division, handed down a decision in the cases of *The Anderson Distilling Company v. Anderson Distilling Company* and *The Anderson Distilling Company v. The Nelson County*

Distilling Company. Judge Toney stated that the defendants Charles Lemon and D. Meschendorf and the Old Times Distillery Company were "perpetually enjoined and restrained" from manufacturing any whiskey under the name of the Anderson Distilling Company or under the name Anderson or from using said names or brands on whiskey. Lemon and Meschendorf were also banned from issuing warehouse receipts for whiskey that was sold under the name of Anderson Distilling Company or Anderson Company. They were banned from delivering whiskey under the name of Anderson Distilling or Anderson Company to anyone and from delivering warehouse receipts that had the name Anderson Distilling or Anderson Company. They were also banned from manufacturing, branding and selling whiskies that had the name Anderson Distilling or Anderson Company. The judge further ruled that Lemon and Meschendorf were banned from selling any whiskey that had the brands or trade names of Anderson Distilling Company or Anderson whiskey.

The same ruling was handed down against the Nelson Distillery Company. The defendants in the Nelson County Distillery were George Mencke, Fred Cornet and H.H. Brueggemann. The Anderson Distilling Company alleged that "Anderson Distilling Company" (under H.H. Brueggemann, John Collins and A.F. Brueggemann) and the Old Times Distillery Company (under Charles Lemon and D. Meschendorf) had worked together and conspired to take advantage of and to pirate and deprive the legitimate Anderson Distilling Company of its right and title to and enjoyment of its trademark, trade name and property. Anderson Distilling Company wanted to stop the Old Times Distillery Company under Lemon and Meschendorf and the Anderson Distilling Company under H.H. Brueggemann, Collins and A.F. Brueggemann from making an inferior quality of whiskey under the name of Anderson Distilling Company and trying to pass off their product on the market as "Anderson" whiskey.

The Anderson Distilling Company alleged that the Nelson Distilling Company conspired for the purpose of having an inferior quality of whiskey made under the name of the Anderson Distilling Company. Anderson stated that the name Nelson Distilling Company would confuse the consumer and that the whiskey would be known as "Nelson" whiskey and traded off as if made by the legitimate Anderson Distilling Company.[347] In 1899, the Old Times Distillery was sold to Ferdinand Westheimer & Sons, wholesale liquor dealers and distillers in Cincinnati, Ohio.

In 1897, the Pleasure Ridge Park Distillery Company went bankrupt, and the plant was sold, with the proceeds going into a trust fund held by

Meschendorf and Lemon. The Mill Creek Distilling Company of Ohio sued Meschendorf for twenty-four barrels of whiskey and $600 in damages for breach of contract.[348]

In 1898, Meschendorf, along with other associates, bought the old Mayflower Distillery, founded in 1880 by H.A. Thierman, and renamed the plant the Old Kentucky Distillery. The main plant was located on the workhouse road, just back of Cherokee Park at Daisy Lane. Meschendorf was one of the most active in the organization of the big distillery, and when the board of directors had its first meeting, the members unanimously elected him president of the company. In 1901, the Old Kentucky Distillery was incorporated with a capital stock of $300,000. The incorporators were D. Meschendorf, O.H. Irvine and D.B. Musselman. O.H. Irvine had formerly worked for the Bonnie Brothers distillery. The flagship brand of Old Kentucky Distillery was Kentucky Dew.[349]

In 1899, the *Louisville Courier-Journal* newspaper reported that Meschendorf bought several hundred barrels of whiskey from the Fible & Crabb Distillery at Eminence, Kentucky, at auction.[350] According to the same newspaper, Meschendorf was also a member of the board of directors for the Beattyville Mineral and Timber Company.

Meschendorf's Kentucky Dew advertisement, 1903. *From* Wine and Spirit Bulletin *(1903)*.

Meschendorf was on the executive committee of the Kentucky Wholesale Liquor Dealers Association and a member of the National Model License League. In 1901, an important meeting took place between the independent distillers and the Kentucky Distilleries and Warehouse Company. They all agreed that the output of bourbon should not exceed twenty-five thousand gallons, in order to keep prices at a cost to where the whiskey men could make a profit. If the market were flooded with bourbon, prices would fall. In 1907, the Kentucky Distillers Association met at the Galt House in Louisville to discuss the different provisions of the pure food law. The members discussed the Food and Drug Act of June 30, 1906, as the law affected the distillery business. Meschendorf was president of the association. The battle was between the blenders or rectifiers and the "straight" bourbon distillers. Meschendorf believed that "straight" Kentucky whiskey distillers should be allowed into the organization and not the blenders or rectifiers, as he believed they did not produce straight bourbon. He agreed with the position of James Wilson, secretary of agriculture, who believed that blended products did not adhere to the pure food law.[351]

In 1901, Meschendorf and S.E. Medley, of Bardstown, bought out the old Davies County Distillery Company, with the main plant and offices located in Owensboro, Kentucky. The Davies Plant was formerly the old R. Monarch plant, which was one of the largest distilleries in the western district. In 1899, the Kentucky Distilleries and Warehouse Company made an effort to acquire the property, with the intention of closing the distillery down permanently, but Mr. Monarch refused to sell. Subsequently, he went bankrupt, and the distillery was run by a trust house under Cowan and Edelen.[352] Meschendorf invested his own money in the Davies County Distillery. The president of the distillery was R.H. Edelen, the vice-president was George G. Brown and secretary and treasurer was G. McGowan. Meschendorf owned 286 shares in the company, Edelen owned 274 shares and Brown owned 228 shares. The distillery had the capacity of three thousand to five thousand barrels annually. Meschendorf paid $40,000 for the plant. The plant had a capital stock of $100,000.[353]

In 1900, Meschendorf married Clara Martin, daughter of a Louisville pioneer jeweler. They had no children. In 1907, Christina Meschendorf died of old age at 318 East College Street, the home of George Martin, who was Dietrich's brother-in-law. She was eighty-one years old.[354] In 1908, Meschendorf purchased the old Karl Jungbluth home in Crestwood, near Beards Station. Karl Jungbluth was president of the McAndrews & Forbes Tobacco Company and built his home in 1888. The thirty-acre estate was a successful Thoroughbred operation owned by Middleton and Jungbluth. The Waldeck stud farm raised the 1903 American Derby winner The Picket. Former winners also included Whirlway, Citation and Swaps. When Meschendorf bought the home, he also bought the breeding farm for Thoroughbreds. His home was one of the most beautiful structures in the neighborhood. By 1911, he had four yearling colts. At Churchill Downs, he had a three-year filly named Dixie Hart. His best horse was George Hermann.[355]

Meschendorf had one brother, Herman Meschendorf, who was a successful farmer in Vicksburg, Mississippi. He also had three nieces and nephews. In October 1911, Dietrich left for San Antonio, Texas, on board a special train hoping that the southern climate would help his failing health. One day after his arrival, he contracted uremic poisoning. A sudden change in his health occurred, and his brother-in-law, William Martin, a jeweler, left for San Antonio. Martin was in New Orleans when a telegram arrived stating that Dietrich had a heart attack and died. He died on November 4, 1911. Martin met the body in New Orleans and brought it back to Louisville for burial.[356] Meschendorf was interred at Cave Hill Cemetery, Section 1, Lot 177, Grave 4. At the time he died, his estate was worth $900,000 (equivalent to $27,409,445 in 2019 dollars). He left one-fourth of his estate to charities and churches, which totaled about $200,000, one of the largest charitable bequests ever made in Louisville.

Meschendorf also approved a monument that would be erected at Cave Hill Cemetery. He left to his widow, Clara, his home in Oldham County, known as Waldeck; the livestock; and half of his estate. One-fourth of his estate was left to relatives, and one-fourth was left to churches and charities. He gave $50,000 to the Children's Free Hospital, $50,000 to the German Evangelical Missouri College and $50,000 to the Louisville Protestant Altenheim. He bought his brother, Herman, and his family a $4,000 home. His nephew, William Meschendorf, was left $6,000, and he also gave his mother, Emelia Meschendorf, $6,000.[357] In 1912, Clara Meschendorf had to pay the sheriff of Oldham County $16,300 in inheritance tax on the estate in Oldham County (equivalent to $424,767 in 2019 dollars).[358]

According to his will, Meschendorf's stock on the Old Kentucky Distillery and Davies County Distillery were to be sold and distributed among his legatees. The Fidelity and Trust Company was the executor and trustee of his will.[359]

Several days after Meschendorf's death, a bonded warehouse of the Davies County Distilling Company, containing more than twelve thousand

Dietrich Meschendorf monument, Cave Hill Cemetery. *Photo by author.*

barrels of whiskey, burned to the ground, a loss of $400,000 (equivalent to $10,643,200 in 2019 dollars). The firemen and volunteers were able to keep the fire contained to one warehouse and saved a nearby warehouse. The burning whiskey flowed into a creek and then into the Ohio River. The Davies County Distillery was owned by Dietrich and Thomas Medley of Owensboro. Medley had died a year before Meschendorf in 1910. The distillery was operated by the Medley brothers.[360] Prohibition put an end to both the Davies County Distillery and the Old Kentucky Distillery.

CHAPTER 24

MAJOR W.H. THOMAS AND PERCY THOMAS

Thomas & Son

William H. Thomas was born in Alexandria, Virginia, and most of his life was spent in Washington, D.C. On June 21, 1848, he married Catherine L. "Kate" Pumphrey in Washington, D.C. Born on February 12, 1829, Catherine was the daughter of Levi and Sarah Miller Pumphrey, residents of Prince George County, Maryland. The Millers were a prominent family in Maryland. Jacob Miller, a brother of Sarah, was one of the leading officers in the Baltimore and Ohio Railroad. Dr. Adam Miller, another brother, practiced medicine in Washington, D.C., but died early, and his widow married Charles Burks, the actor, who was a half-brother of Joseph Jefferson.[361]

When the Civil War broke out, William and Catherine were living in Washington, D.C., and William joined the Confederate army. She decided to pack her trunks, and when he told her goodbye, she announced her intention to go with him. She followed her husband's fortunes during the Civil War, first with the Army of Northern Virginia and later in the West. She was well known to most of the prominent Virginia officers, including Confederate general Robert E. Lee. The most cherished piece of the Civil War she was able to preserve was an original order of General Lee dated October 19, 1861. When the Battle of Richmond, Kentucky, was fought, Catherine Thomas was in Knoxville, Tennessee, and the letter received from her husband giving an account of the battle was telegraphed in substance to President Jefferson Davis; it was the first report received at the Confederate capital.[362]

During the Civil War, William Thomas rose to the rank of major and served on the staff of Confederate general Edmund Kirby Smith. During the Confederate invasion of Kentucky in 1862, he came to Kentucky with Confederate general Braxton Bragg's Army of Mississippi; during the invasion, he was stationed in Lexington, Kentucky.[363]

Percy Thomas, the son of William and Catherine, was born in 1852 in Washington, D.C., and received his early education in Washington. When Percy was very young, he and his mother moved with his father to a number of different cities during the Civil War in an effort to be as near to him as possible. At the close of the Civil War, the family moved to Lexington, and Percy received his college education in that city, graduating from Kentucky State University. In 1872, he came to Louisville and entered the firm of Newcomb & Buchanan, whiskey distillers. He continued to work for the company until 1879, when the company closed.[364]

Shortly after the closure of the firm, Percy and his father formed W.H. Thomas & Son. During the 1880s and 1890s, Major Thomas and his son owned one of the largest whiskey establishments in Louisville, located on Main Street. Major Thomas was not a distiller, but more of a holder of fine whiskies. Whenever he heard of a lot of fine whiskey up for auction, he would buy the lot as soon as possible and then sell the whiskey to wholesale dealers. At one time, he was rated as being worth at least $1 million. For thirty years, Major Thomas and Percy held the record as the foremost holders of fine and straight whiskies in the country. He possessed some of the finest whiskies in the world.[365]

In 1892, the market began to change, and in order to handle his fine whiskies at the least expense possible, Major Thomas exported most of them, keeping the whiskey in warehouses in Bremen and other places. Bringing his whiskey back to the country in 1893, when the panic came on, was disastrous to a large extent. Most of the goods were forced out of bond and with no market. About this time, the rectifiers began to flourish, and a few years later, the Whiskey Trust was formed, which was detrimental to many of the independent dealers. Major Thomas sold off his straight whiskies as best as he could, but the tide was against him. In April 1893, creditors of W.H. Thomas and Son were called together, and on May 18, they met at the Clearing House. A statement of the financial status of the firm was prepared for the creditors and showed that the company had liabilities of $1 million, with assets amounting to $1,150,000, making a surplus of $150,000. The creditors allowed an extension of one to two years to realize on the whiskey. The market did not improve, and Thomas could not realize on the stock of whiskey he had on hand. Thomas managed to pay off $400,000 of their obligations, but they still owned $600,000. Nearly all of money secured was in pledges of whiskey. During the Panic of 1893, new whiskey was much easier to sell than old because the new whiskey, not having been taken out of bond, could be sold upon the prospects of better times by the expiration of the three-year bond period, whereas the older whiskey, having been freed, could hardly sell at all. While the whiskey was in warehouses, a considerable loss had to be suffered by Thomas & Son from the evaporation.

Thomas & Son also owned a large interest in the J.G. Mattingly Company, whose stock had been held by the Kentucky National Bank as collateral. The bank decided to sell the stock to the highest bidder to satisfy the claims and advertised that the buildings were to be sold at auction. There were 499 shares of stock up for auction. Major Thomas realized the great blow that the disposal of the stock would have on him, and he sued to stop the sale. The judge in the case decided not to grant the injunction. Major Thomas had no choice but to file a deed of assignment. Thomas also sued Paul Jones, president of J.G. Mattingly Company, asking for the appointment of a receiver and declaring of a dividend on $300,000 profits. The Louisville Trust Company, the assignee, promised that the creditors would be paid.[366]

Thomas & Son managed to survive for a few more years, but in 1901, William and his son filed a petition for bankruptcy, with liabilities amounting to $500,532. No assets were listed. W.H. Thomas owed $23,256 in secured claims and $40,094 in unsecured claims. The firm's assets were mostly in real estate, which was in the hands of the Louisville Trust Company. They also owned hundreds of barrels of whiskey, including Old Jordan and Brook and Harris (325 barrels); Anderson and Nelson County Distillery (205); and Bartley, Johnson and Company (50 barrels), which they held as securities with different banks—for example, the Fourth National Bank gave $11,485 to Thomas & Son, which was secured by whiskey valued at $12,000, and Lawrence Bank gave Thomas and Son $24,062, which was secured by 663 barrels of W.B. Saffel whiskey valued at $25,000.[367]

On October 26, 1905, Catherine Thomas died of paralysis at the family home on 908 Fourth Street. Her funeral was held at Christ Church Cathedral. Her burial took place at Cave Hill Cemetery, where she is buried, Section P, Lot 860, Grave 2.

All through the years of financial trouble, Major Thomas hoped that the day would come when the fine straight whiskies would make a comeback. In 1906, the Pure Food and Drug Law was passed, and the federal government put an end to rectifiers or blenders. But the law came too late for Major Thomas. His wealth was gone, and his life was coming to an end. He was never able to adapt to the new methods or business systems. He spent most of his time in the Louisville hotels and was an excellent conversationalist. He was always ready with a store of reminiscences. He loved to talk about the Civil War and could relate many interesting incidents pertaining to the war. He took a great interest in the fight between the whiskey interests when the Pure Food Law was under discussion at Washington, D.C., and all over the country.[368] He was a stickler for the name *whiskey* to be applied only to the fine straight product. He even testified before Congress at a Senate committee hearing. He stated that "the whiskey that moonshiners sell is more honestly made than what these compounders and blenders make."[369]

In 1907, a deed of trust filed by W.H. Thomas & Son Company to the Louisville Trust Company resulted in the passing of one of the oldest and best-known wholesale whiskey firms in the South. The purpose of the deed was to wind up the affairs of the company. Hector Loving, president of the Louisville Trust Company, stated that the company would be liquidated in eight months. The company owned 3,700 barrels of old whiskey, which was practically the entire supply of old whiskey in the country. Loving stated that the old whiskey owned and controlled by the company included most of the famous brands distilled. Thomas stated that the step was necessary because he had difficulty borrowing money from the banks on old whiskey.[370]

On October 6, 1908, Major Thomas died of old age at St. Anthony's Hospital. His funeral was held at Pearson's Funeral Home on Third and Chestnut. He was buried at Cave Hill Cemetery, Section P, Lot 860, Grave 3.[371]

In 1883, Percy Thomas married Mary Patrick. She was a native of Pennsylvania. They lived for a while in Shelbyville, but in order to be closer to his business, he moved to Louisville. For a number of years, they lived in the Galt House, but they later moved in with A.P. Speed and family on Park Avenue. They had two children: Katherine Thomas and Mrs. Ruth Thomas, who married Philip Weissinger. They also had one grandson, Harry Weissinger. On January 23, 1911, while sitting at the dinner table at his home on 417 Park Avenue, Percy suffered a heart attack and died immediately. He was fifty-seven years old. His body was moved to the home of W.H. May on 1209 South Third Street, and funeral services were conducted. He was buried at Cave Hill Cemetery, Section P, Lot 860, Grave 4.[372]

In April 1911, Mary Patrick Thomas, Percy Thomas's wife, decided to sell the Percy Thomas Art Collection. The collection was valued at $150,000 (equivalent to $3,964,844 in 2019 dollars), and during the lifetime of Major William Thomas, the art gallery was visited by art connoisseurs from every section of the country. The collection consisted of seventy-five paintings.[373]

CHAPTER 25

DAVID SACHS & SONS

Marion County Distillery and Puritan Rye

David Sachs was a native of Germany and arrived in America during the 1840s, settling initially in Springfield, Illinois. His sons, Morris and Edward, were both born in Springfield. In 1872, he came to Louisville and founded the wholesale liquor and distillers firm of D. Sachs.[374]

In 1882, his two sons joined his company, and the name was changed to David Sachs & Sons. His flagship brand was Puritan Rye. His advertisements stated, "'Kick' If You Don't Get Puritan Rye" and "Its All Right." His advertisements also noted that David Sachs & Sons was the sole controller of Registered Distillery No. 11—5th District Kentucky, Marion County, Kentucky, with its main office in the Board of Trade building on 152 Third Street in Louisville, Kentucky. Sachs & Sons also had the image of a Puritan elder on its label. Sachs & Sons controlled the output of a number of distilleries and made prominent specialties of the celebrated Oakland and Delmonico bourbon and sour mash whiskies. It also handled the leading brands of Kentucky whiskies.[375] David was president of the Hebrew Relief Society and a prominent member of the Temple Adath Israel. He was the brother-in-law of Julius and Louis Barkhouse.[376] On June 6, 1898, David died of a fatal stroke at his home on 1423 Second Street. He had been at the bedside of his wife, who was seriously ill. She died one day later. On June 8, 1898, the funeral services were held for David and Hannah Sachs at their home on Second Street, and both were buried side by side at the Adath Israel Cemetery. He was survived by his sons, Morris and Edward Sachs, and his three daughters: Mrs. Julie Sachs, married to Sam Haas; M.A. Sachs; and Mrs. Bertha Sachs, married to Ben Nahm.[377] After his death, his sons continued to run the company for eleven years. Sam Haas also became a member of the company. In 1919, Prohibition led to the closure of David Sachs & Sons.

On September 10, 1923, Sam Haas died at the age of sixty at Johns Hopkins Hospital in Baltimore, Maryland. His funeral was held at his home at 409 Kensington Court in Louisville. Haas was born in New Orleans on January 16, 1863. He was a Shriner, a trustee of the Jewish Welfare Federation, a member of the Standard Club, a member of the board of presidents of the Welfare League, a member of the Waverly Hills Tuberculosis Board and a member of the Temple Adath Israel Congregation. He was survived by his widow, Julia Sachs Haas. He was buried at the Adath Israel Cemetery in Louisville.[378]

On November 6, 1932, Morris D. Sachs died at his apartment at the Brown Hotel. He was eighty years old. He was originally from Springfield, Illinois, and came to Louisville at the age of twenty to enter the liquor business with his father. For forty years, he was a member of David Sachs & Sons. He was a charter member of the Standard Club, Adath Israel Congregation and B'nai B'rith. He was survived by his brother, Edward Sachs, and sister, Julia Sachs Haas. He was buried at Adath Israel Cemetery in Louisville, Kentucky.[379] When he died, he left an estate valued at $150,000. He left $7,950 to different charities. His estate was to be divided between Julia Haas, Mrs. Cecil Sachs Heyburn, Maude Oppenheimer, Florence Blum, Walter Nahm and brothers-in-law Victor, David and Isaac. He left money to the Jewish Orphans Home, Sir Moses Monteflore Home, the National Hospital for Consumptives Denver, the Jewish Hospital and the Jewish Welfare Federation. He also left money to the Jewish Children's Home, Children's Free Hospital and Congregation Adath Israel. He gave money to the Young Men's Hebrew Association, the Susan Speed Davis Home and the Pine Mountain Settlement.[380]

On September 21, 1938, Edward Sachs died at his home at Commodore Apartments and left an estate of $70,000 to his wife, Clara Sachs. He came to Louisville in 1872 and entered the wholesale whiskey business with his father and brother. After Prohibition closed the doors of David Sachs & Son, Edward became president of the Louisville Realization Company and a director of the Lincoln Bank and Trust Company. He was also once one of the leaders of the Community Chest. He was a member of the Adath Israel Congregation of B'nai B'rith Lodge and of the Young Men's Hebrew Association. He was survived by his wife, Clara, and two daughters, Edith Grabfelder and Katherine Sachs. His body was cremated.[381]

NOTES

Introduction

1. Johnston, *Memorial History of Louisville*, 261.
2. Ibid.
3. Elijah Craig website.
4. Johnston, *Memorial History of Louisville*, 262.
5. Huckelbridge, *Bourbon*, 64.
6. Ibid., 118.
7. Ibid., 119.
8. Ibid., 122.
9. Ibid., 126.
10. Allison, *City of Louisville and a Glimpse of Kentucky*, 16.
11. *Louisville: Nineteen Hundred and Five*, 18.

Chapter 1

12. "Former Prince of Sports Dies, Col. 'Jim' Douglas Succumbs to Age and Disease," *Louisville Courier-Journal*, January 3, 1917, 1.
13. Ibid.
14. *Douglas v. Kentucky*, November 29, 1897, 168 U.S. 488 (1897).
15. Ibid.
16. Ibid.

17. *Wine and Spirit Bulletin* 17 (1903).
18. "Newsy Notes of Trotters, Louisville Becoming the Center for Breeding Farms," *Louisville Courier-Journal*, November 7, 1898, 6.
19. "Big Deal in Horses, John E. Madden Buys All the Eothen Colts and Fillies from Col. J.J. Douglas, of This City," *Louisville Courier-Journal*, March 17, 1899, 6.
20. "Former Prince of Sports Dies," 1.
21. Talbott, "Douglas Park Racetrack."
22. "Douglas Horses Sold: Pilatus Brings $1,800, while Mayme Nutwood Sells for $340," *Louisville Courier-Journal*, October 4, 1905, 7.
23. "Turf War in Louisville, Western Jockey Club Secures Douglas Park Track," *Louisville Courier-Journal*, November 29, 1905, 1.
24. "Close Deal for Douglas Track: Money Paid and Property Transferred to Louis A. Cella," *Louisville Courier-Journal*, December 10, 1905, C4.
25. "Big Turf Deal Goes Through: Owners of Churchill Downs and Douglas Park to Pool Interests," *Louisville Courier-Journal*, February 26, 1907, 8.
26. "Former Prince of Sports Dies," 1.
27. Ibid.
28. Ibid.
29. "For Creditors Committee Takes Over Affairs of Distilling Companies," *Louisville Courier-Journal*, September 17, 1915; "$70,000 for Creditors of J.J. Douglas Company," *Louisville Courier-Journal*, October 17, 1916, 10.
30. "Former Prince of Sports Dies," 1.
31. "Douglas Will Contest Ends," *Louisville Courier-Journal*, August 29, 1917, 10.
32. Kleber, *Encyclopedia of Louisville*, 251.

Chapter 2

33. "Stricken: While Sitting in His Arm Chair Thomas H. Sherley, Dead Suddenly Seized with Paralysis of the Heart Was a Foremost Distiller: A Sturdy and Enterprising Citizen Who Accomplished Much for the Public Good: Some of His Characteristics," *Louisville Courier-Journal*, November 30, 1898, 4.
34. Ibid.
35. Ibid.
36. Ibid.
37. Ibid.
38. Ibid.

39. Ibid.
40. Ibid.
41. Ibid.
42. Ibid.
43. Ibid.
44. Ibid.
45. Ibid.
46. Ibid.
47. Ibid.
48. Ibid.
49. Ibid.
50. "The Last Tribute: Funeral of Mr. Sherley Will Take Place Today," *Louisville Courier-Journal*, December 1, 1898, 10.
51. "The Last Tribute: Funeral of Mr. Sherley from the Cathedral," *Louisville Courier-Journal*, December 2, 1898, 16.

Chapter 3

52. "Illness Fatal to John M. Atherton, 91," *Louisville Courier-Journal*, June 6, 1932, 1; *Industries of Louisville and New Albany, Indiana*, 93.
53. "Illness Fatal to John M. Atherton, 91," 1; *Industries of Louisville and New Albany, Indiana*, 93.
54. Howell and Waters, *Hardin and LaRue Counties*, 11.
55. "Illness Fatal to John M. Atherton, 91," 1.
56. *Southwestern Reporter* 9 (August 6, 1888–January 7, 1889): 520.
57. Ibid.
58. Ibid.
59. "Deal Closed: Trust Takes Atherton and Patterson Whiskey Was Long Pending Over a Million Dollars Involved in the Sales Mr. Atherton Plans," *Louisville Courier-Journal*, June 4, 1899, A10.
60. "Illness Fatal to John M. Atherton, 91," 1.
61. Ibid.
62. Ibid.
63. Ibid.
64. Ibid.
65. "Mr. Atherton's Gift: Letter Explaining His Donation to Georgetown Baptist College Believes in Paying Good Wages and Wants the Teacher's Salary Raised," *Louisville Courier-Journal*, February 3, 1893, 6.

66. "Illness Fatal to John M. Atherton, 91," 1.
67. Ibid.
68. Ibid.

Chapter 4

69. *Memoirs of the Lower Ohio Valley*, 286.
70. Ibid.
71. "Papers Ready: Grabfelder to Buy Murphy-Barber Distillery to Be Put in Operation, Price Said to Be Between $30,000 and $40,000," *Louisville Courier-Journal*, September 27, 1899, 9.
72. "Keep Open House to Say Farewell: Mr. and Mrs. Samuel Grabfelder Receive Handsomely at the Standard Club," *Louisville Courier-Journal*, March 31, 1905, 4; "Grabfelder Home Sold Yesterday to Mrs. W.H. Kaye, Sr.," *Louisville Courier-Journal*, January 9, 1905, A3.
73. "Sam'l Grabfelder, Millionaire Dead, Man Who Made Fortune Here Was Known for Charity," *Louisville Courier-Journal*, April 18, 1920, B1.
74. "Welfare Bodies Get $10,000 Each: Jewish Charity and Hospital Benefit by Bequests of Samuel Grabfelder," *Louisville Courier-Journal*, January 18, 1922, 14.
75. "$20,000 Bequeathed to Jewish Bodies Here, Estate of Samuel Grabfelder Former Louisville Man, Is Estimated at $6,000,000," *Louisville Courier-Journal*, May 8, 1920, 1.
76. "Grabfelder to Be Buried Today," *Louisville Courier-Journal*, May 19, 1924, 14.

Chapter 5

77. "Death of D.H. Taylor: Severe Cold Unexpectedly Develops into Pneumonia and the End Speedily Follows," *Louisville Courier-Journal*, March 17, 1894.
78. Ibid.
79. "Donald McWain, Odd Lots, Brand Name Crowds Producing Firm into Background," *Louisville Courier-Journal*, December 2, 1944, 12.
80. "Succumbs to Nervous Trouble After a Long Illness, Mr. J.T. Williams Is Dead," *Louisville Courier-Journal*, February 23, 1903, 2.
81. "Mr. Williams Funeral This Afternoon, Remains Will Be Placed in Vault and Taken to Nashville for Burial," *Louisville Courier-Journal*, February 24, 1903, 8.

82. "Death of D.H. Taylor."
83. "Mr. Williams Funeral This Afternoon," 8.
84. "Compromise, Will Probably Effected Among J.T. Williams Heirs," *Louisville Courier-Journal*, August 9, 1903, 4.
85. Ibid.
86. "Employees Have Interest in Firm of Taylor and Williams, Will Be Incorporated," *Louisville Courier-Journal*, March 11, 1903, 6.
87. "Donald McWain, Odd Lots, Fidelity Trust and Columbia Seeks Distillery Bids by Noon Saturday," *Louisville Courier-Journal*, November 9, 1943, 16.
88. "Donald McWain, Glenmore Purchases Distillery Here, More than Five Million Paid for Taylor & Williams Plant, Whiskey Stock, Other Assets," *Louisville Courier-Journal*, April 15, 1944, 1.

Chapter 6

89. "Death Claims C.P. Moorman," *Louisville Courier-Journal*, February 14, 1917, 1.
90. Ibid.
91. Ibid.
92. "Mr. Semple Become Vice President, Stockholders of C.P. Moorman & Co., in Which He Purchased Interest Meet Today," *Louisville Courier-Journal*, July 7, 1903, 14.
93. "Louisville Distillers Buy Outside Property, Moorman & Co. Acquire Samuels Plant," *Louisville Courier-Journal*, January 30, 1912, 10.
94. "Death Claims C.P. Moorman," 1.
95. "Moorman Out: Sells Holdings of Fidelity Stock for $237,000," *Louisville Courier-Journal*, February 7, 1912, 6.
96. "Death Claims C.P. Moorman," 1.
97. Ibid.
98. "Huge Bequest Made Charity: Moorman Will Provide Home for Aged Women," *Louisville Courier-Journal*, February 19, 1917, 1.
99. "Baby Born to Mrs. Bartlett Ends Fight Over Moorman Estate Worth Millions," *Louisville Courier-Journal*, August 18, 1920, 1.
100. "Beauty and Comfort Grace C.P. Moorman Home for Indigents," *Louisville Courier-Journal*, February 23, 1930, 71.
101. Hotaling & Company website.

Chapter 7

102. "Varied Career: Brought to Close in John G. Roach's Death, Soldier, Distiller, Politician, and Businessman, Taken Suddenly Ill of Apoplexy on Thursday, Second Attack, on Sunday," *Louisville Courier-Journal*, December 10, 1907, 3.
103. Ibid.
104. Ibid.
105. "Big Plant: Combine Acquires Roach Distillery Deal Closed Yesterday Whiskey on Hand Held by Mr. Roach," *Louisville Courier-Journal*, March 15, 1899, 8.
106. "Burning Whiskey," *Louisville Courier-Journal*, November 30, 1906, 2.
107. "Big Plant," 8.
108. Ibid.
109. "John G. Roach's Will," *Louisville Courier-Journal*, February 9, 1908, 12.

Chapter 8

110. "Paul Jones Funeral: Remains Deposited in a Vault in Cave Hill," *Louisville Courier-Journal*, February 27, 1895.
111. Four Roses Bourbon, "Four Roses History and Heritage."
112. "Paul Jones Funeral."
113. Ibid.
114. Ibid.
115. Four Roses Bourbon, "Four Roses History and Heritage."
116. "Paul Jones: Death Comes Suddenly to the Wealthy Distiller," *Louisville Courier-Journal*, February 24, 1895.
117. Ibid.
118. Reigler, *Kentucky Bourbon Country*, 117; Kleber, *Encyclopedia of Louisville*, 453.

Chapter 9

119. "The Phil Hollenback [*sic*] Co. Move into Larger Quarters: Firm Stands in Front of the Great Industrial Enterprises of the Country," *Louisville Courier-Journal*, March 26, 1911, B7.
120. Ibid.

121. Ibid.
122. Ibid.
123. Ibid.
124. Ibid.
125. Ibid.
126. Ibid.
127. Ibid.
128. Ibid.
129. "Glencoe Plant Operating at Full Capacity," *Louisville Courier-Journal*, January 1, 1937, 8, Section 3.
130. "Glencoe Distillery Is Sold to National," *Louisville Courier-Journal*, April 8, 1943, 1.
131. "Ex-Head of Distillery L.J. Hollenbach, Dies," *Louisville Courier-Journal*, October 15, 1954, 15.

Chapter 10

132. "Jas. Thompson Called by Death, Head of Distillery Firm Dies at Home in Anchorage," *Louisville Courier-Journal*, December 12, 1924, 1–3.
133. Ibid.
134. Kleber, *Encyclopedia of Louisville*, 431.
135. "Jas. Thompson Called by Death," 1–3.
136. Ibid.
137. Ibid.
138. "Thompson Leaves $335,000 Estate," *Louisville Courier-Journal*, December 19, 1924, 24.
139. "Frank Thompson, Ex-Chief of Glenmore Distilleries Dies," *Louisville Courier-Journal*, February 28, 1990, B6.
140. "Col. Frank B. Thompson Is Just as He Appears in Those Whiskey Ads," *Louisville Courier-Journal*, August 21, 1960, 70.
141. Ibid.
142. Ibid.
143. "Frank Thompson Jr., Former Head of Glenmore Distilleries Co., Dies," *Louisville Courier-Journal*, May 14, 1975, 35.
144. Ibid.
145. "Frank Thompson, Ex-Chief of Glenmore Distilleries, Dies," 7.
146. "Susan Tompor, Glenmore Strengthens Ties to Liquor Industry," *Louisville Courier-Journal*, October 24, 1988, 17.

147. "Judith Egerton, Glenmore Is Sold to British Distillery," *Louisville Courier-Journal*, July 12, 1991, 1.

Chapter 11

148. "Nathan Block Passes Away: Retired Whiskey Broker Since 1870," *Louisville Courier-Journal*, January 2, 1917, 2.
149. *Industries of Louisville and New Albany, Indiana*, 84.
150. Ibid., 84.
151. "Nathan Block Passes Away," 2.
152. Ibid.
153. Ibid.
154. "Nathan Block Home Sale Price $20,000," *Louisville Courier-Journal*, June 12, 1907, 6.
155. "Nathan Block Passes Away," 2.
156. "Children Get Block Estate: Will Disposing of $100,000 Is Probated," *Louisville Courier-Journal*, January 7, 1917, B10.
157. "Before Mirror: Joseph C. Block Stands to Shoot Himself," *Louisville Courier-Journal*, January 26, 1908, 8.
158. "Bernard N. Block," *Louisville Courier-Journal*, September 4, 1966, A17.
159. "W.S. Block Despondent, Ends Life with Bullet," *Louisville Courier-Journal*, May 22, 1915, 12.

Chapter 12

160. "Death Ends Long Suffering of George W. Swearingen," *Louisville Courier-Journal*, December 19, 1901, 4.
161. Ibid.
162. Haara, *Bourbon Justice*, 136.
163. *The Pharmaceutical Era* (January 15, 1891): xii–xiv.
164. *Mellwood Distilling Company v. Harper C.C.*, *Federal Reporter*, vol. 167, April–May 1909.
165. "Death: Ends Long Suffering," 4.
166. Johnston, *Memorial History of Louisville*, 614–15.
167. "Mr. Swearingen Buried in Cave Hill," *Louisville Courier-Journal*, December 20, 1901, 5.

Chapter 13

168. "Gathered to His Fathers: Henry McKenna the Well Known Distiller Dies Suddenly of Heart Disease a Self Made Man Who Came Up from Humble Beginnings, Noted for His Charities, Other Deaths," *Louisville Courier-Journal*, February 23, 1893.
169. "Old McKenna Distillery Reopens, February 18, 1934," *Louisville Courier-Journal*, 55.
170. Ibid.
171. Ibid.
172. "Stafford McKenna Dies," *Louisville Courier-Journal*, January 17, 1935, 10.
173. "J.S. McKenna Head of Distillery at Fairfield Dies," *Louisville Courier-Journal*, December 18, 1941, 32.
174. "Seagram Pays $950,000 for McKenna Distillery," *Louisville Courier-Journal*, September 23, 1941, Sec. 1.

Chapter 14

175. Bernheim, *Reflections of Isaac Wolfe Bernheim.*
176. Ibid.
177. Ibid.
178. Ibid.
179. Ibid.
180. Ibid.
181. Ibid.
182. Ibid.
183. Ibid.
184. Ibid.
185. "Munificent Gift of Messrs. Bernheim to the City," *Louisville Courier-Journal*, September 21, 1899, 1.
186. "New Building: Y.M.H.A. Will Celebrate This Evening Messrs Bernheim Gift," *Louisville Courier-Journal*, January 1, 1896, 7.
187. "Henry Clay and Ephraim McDowell Get Place in Nation's Hall of Fame," *Louisville Courier-Journal*, April 26, 1927, 1.
188. "Howard Hardaway, Bernheim Forest Looks Ahead," *Louisville Courier-Journal*, September 16, 1945, 68.
189. "Louisville Receives as a Gift from Mr. and Mrs. Isaac W. Bernheim an Heroic Statue in Bronze of Abraham Lincoln, the Work of George

Gray Bernard, the Famous Sculptor," *Louisville Courier-Journal*, October 27, 1922, 1.

190. Bernheim, *Reflections of Isaac Wolfe Bernheim*.

191. "I.W. Bernheim Falls to Death in California," *Louisville Courier-Journal*, April 2, 1945, 1.

192. Kleber, *Encyclopedia of Louisville*, 86–87; "Ex-Distillery Official Frank Bernheim Dies," *Louisville Courier-Journal*, August 23, 1962, 13.

Chapter 15

193. "Marion E. Taylor Taken by Death, Retired Distiller Was One of Largest Holders of Real Estate in Louisville," *Louisville Courier-Journal*, July 2, 1921, 1.

194. "Whiskey and Brewing," *Louisville Courier-Journal*, March 13, 1897, A6.

195. "Kentucky Whiskies: A Product in Which This State Is Connected to Lead All Others," *Louisville Courier-Journal*, June 1, 1891, 12.

196. "Trade Marks: The Basis of a Big Suit Filed Yesterday, $150,000 Demanded, E.H. Taylor Jr. & Company Enjoin Marion E. Taylor of Louisville," *Louisville Courier-Journal*, June 10, 1896, 8.

197. "Remanded: Court of Appeals Rules on Taylor Whisky Case: Involved Use of Trade Mark 'Old Kentucky Taylor' on Whisky Put On Market," *Louisville Courier-Journal*, March 18, 1905, 7.

198. *Wine and Spirit Bulletin* 17 (1903): 27.

199. "Remanded: Court of Appeals Rules on Taylor Whiskey Case," 7.

200. Ibid.

201. Haara, *Bourbon Justice*, 136, 102–3.

202. Ibid.

203. Ibid.

204. "To Raise Fund: Louisville Will Help Fever Stricken Natchez," *Louisville Courier-Journal*, October 12, 1905, 8.

205. "Marion Taylor Taken by Death," *Louisville Courier-Journal*, July 2 1921, 1.

206. "Marion E. Taylor Welcomed to Board of Trade," *Louisville Courier-Journal*, October 27, 1910, 10.

207. "Paul Jones Building Sold: Marion E. Taylor Buys Block; $1,000,000 Said to Be Price," *Louisville Courier-Journal*, April 15, 1919, 5.

208. "Marion Taylor Taken by Death," 1.

209. Ibid.

210. "Taylor Estate Tops $1,500,000: Widow Gets Half; Remainder Goes to Kin, Servants, and Christmas Fund, Will Filed for Probate," *Louisville Courier-Journal*, July 6, 1921, 7.

Chapter 16

211. "Death Claims Nathan M. Uri: Succumbs to Bright's Disease After Long Illness," *Louisville Courier-Journal*, February 23, 1909, 1.
212. Ibid.
213. Bernheim, *Reflections of Isaac Wolfe Bernheim.*
214. Ibid.
215. Ibid.
216. "Whiskey and Brewing," A6.
217. "Death Claims Nathan M. Uri," 1.
218. Ibid.
219. Ibid.
220. Ibid.
221. Ibid.
222. Ibid.
223. Ibid.
224. Ibid.
225. "$140,000 Fire on Main Street, Blaze Starts in N.M. Uri Whiskey Establishment," *Louisville Courier-Journal*, April 2, 1915, 1.
226. "Walter Uri, 39, Taken by Death," *Louisville Courier-Journal*, January 16, 1929, 5.
227. "Expires at 53, Morris Uri," *Louisville Courier-Journal*, September 26, 1936, 11.

Chapter 17

228. *Wine and Spirit Bulletin* 18 (1904): 21.
229. "Failure of Heart Disease Causes Robert P. Bonnie's Death," *Louisville Courier-Journal*, January 16, 1904, 4.
230. "W.O. Bonnie Sr. Called by Death," *Louisville Courier-Journal*, March 2, 1923, 1.
231. "End Comes to Frank W. Bonnie, After Eight Days Illness," *Louisville Courier-Journal*, July 30, 1904, 12.

232. Kentucky Court of Appeals, *Bonnie & Company v. Bonnie Brothers*, 169 S.W. Rep. 871, October 13, 1914, *Trade Mark Reporter*, vol. 4, United States Trade-Mark Association, New York, 1914, 548.
233. "End Comes to Frank W. Bonnie," 12; "Bequeaths to Family Estate Worth $200,000 Frank W. Bonnie's Wishes," *Louisville Courier-Journal*, August 4, 1904, 10.
234. "Failure of Heart Disease," 4.
235. "E.S. Bonnie Dead," *Louisville Courier-Journal*, May 16, 1907, 8.
236. Kentucky Court of Appeals, *Bonnie & Company v. Bonnie Brothers*, 548.
237. Ibid., 551.
238. Ibid.
239. Ibid.
240. "W.O. Bonnie Sr. Called by Death," 1.
241. Kleber, *Encyclopedia of Louisville*, 249.

Chapter 18

242. "Kentucky Whiskies," 18.
243. "Men Who Made Money: The New Savings Bank Organized by Its Stockholders," *Louisville Courier-Journal*, March 25, 1891, 5.
244. "Individual Assignments, The Sweetwood Distillery and President Schwartz Forced into Liquidation," *Louisville Courier-Journal*, July 26, 1893, 8.
245. Ibid.
246. "Schwartz Various Liabilities," *Louisville Courier-Journal*, July 31, 1893, 8.
247. Ibid.
248. "The German National Sued," *Louisville Courier-Journal*, September 28, 1893, 6.
249. Ibid.
250. "Whole Thing a Fraud," *Louisville Courier-Journal*, April 5, 1894, 6.
251. "M. Schwartz's Collateral, Stock and Dark Hollow Bonds Disposed of to Satisfy a Judgement," *Louisville Courier-Journal*, May 8, 1894.
252. "Fifty Thousand in Notes," *Louisville Courier-Journal*, March 2, 1894, 8.
253. "No Conspiracy," *Louisville Courier-Journal*, January 19, 1895, 8.
254. "Sharpe Loses: Has No Interest in the Matters Complained of. Schwartz Was Insolvent," *Louisville Courier-Journal*, November 1, 1896, A2.
255. "M'Knight's Story of the Looting of the German National," *Louisville Courier-Journal*, February 19, 1897, 1.

256. "Identified: Hart's Picture as the Man Who Swindled Him, New York Police Thought They Had Moses Schwartz," *Louisville Courier-Journal*, December 23, 1901, 2.
257. Ibid.
258. "Schwartz? Is Philadelphia Suspect Louisville Promotor?," *Louisville Courier-Journal*, March 27, 1902, 4.
259. "Moses Schwartz, Banker Who Went Broke, Now Living in Ease and Affluence in East," *Louisville Courier-Journal*, April 7, 1911, 5.

Chapter 19

260. "Philip Stitzel Dead: Prominent Citizen Succumbs to Heart Disease," *Louisville Courier-Journal*, March 28, 1904, 4.
261. "Racks for Tiering Barrels," Frederick Stitzel, Louisville, Kentucky, #221,945, filed August 22, 1879, Specifications and Drawings of Patents Issued from the U.S. Patent Office, U.S. Patent Office, November 1879, Washington, D.C., 1879.
262. Campbell, *But Always Fine Bourbon*.
263. Ibid.
264. "Philip Stitzel Dead," 4.
265. Campbell, *But Always Fine Bourbon*.
266. "W.L. Weller Dead," *Louisville Courier-Journal*, March 24, 1899, 2.
267. "Destroyed by Fire, Burning of W.L. Weller & Sons Redistilling Establishment Last Night," *Louisville Courier-Journal*, March 14, 1873, 4.
268. Ibid.
269. Campbell, *But Always Fine Bourbon*.
270. "W.L. Weller Dead," 2.
271. "W. Weller & Sons Incorporated," *Louisville Courier-Journal*, January 1, 1908, 7.
272. "Eight Men Hurt Fighting Fire, Main Street Whisky House Goes Up in Flames," *Louisville Courier-Journal*, August 29, 1909, C1.
273. "Jacob Stitzel, Retired Distiller, Dies at 65 Years," *Louisville Courier-Journal*, November 21, 1913, 4.
274. "Merger of Old Whisky Houses: Weller and Trost Bros. United $250,000 Assets," *Louisville Courier-Journal*, February 23, 1915, 2.
275. "Deaths and Funerals: John C. Weller," *Louisville Courier-Journal*, April 1, 1918, 7.

276. "75,000 Gallons Daily Capacity of Distillery, Stitzel Weller Plant Occupies 20-Acre Site at Shively," *Louisville Courier-Journal*, January 1, 1937, 6.
277. Campbell, *But Always Fine Bourbon*.
278. "75,000 Gallons Daily Capacity of Distillery," 6.
279. Ibid.
280. Ibid.
281. Mitenbuler, *Bourbon Empire*, 109.
282. Tamazzo, *Bullets and Bourbon*.
283. "Ex-Distiller Is Taken by Death, Frederick Stitzel, 81, Was Inventor of Railroad Semaphore," *Louisville Courier-Journal*, September 19, 1924, 18.
284. "J.P. Van Winkle, Dean of All Distillers, Still Thinks the Old Methods Are the Best," *Louisville Courier-Journal*, August 19, 1956, 4.
285. "Arthur Philip Stitzel, 72, Official of Distillery Here, Dies," *Louisville Courier-Journal*, April 14, 1947, 4.
286. "Stitzel-Weller, Distillery Executive Van Winkle Dies," *Louisville Courier-Journal*, February 17, 1965, 1.
287. Heaven Hill Distillery, "Old Fitzgerald."
288. Bulleit Frontier Whiskey, "Stitzel Weller Distillery."
289. "Julian Van Winkle Dies, Was Chief of Distilleries," *Louisville Courier-Journal*, November 21, 1981, 15.
290. Campbell, *But Always Fine Bourbon*.
291. Chuck Martin, "Bottling a Bourbon Legend," *Cincinnati Enquirer*, April 30, 2000, H1.
292. Old Rip Van Winkle, "Heritage."

Chapter 20

293. "Death Comes: J.T.S. Brown Dies After Severe Illness," *Louisville Courier-Journal*, October 18, 1905, 10.
294. "J.T.S. Brown Jr. Dies at Home in Florida," *Louisville Courier-Journal*, December 7, 1940, 8.
295. Ibid.
296. Ibid.
297. "J.T.S. Brown & Sons Close the Deal for the Waterfill Distillery," *Louisville Courier-Journal*, December 28, 1894, 8.
298. "Large Estate, J.T.S. Brown Sr. Left Property to Children," *Louisville Courier-Journal*, October 28, 1905, 7.

299. "J.T.S. Brown & Sons Charter," *Louisville Courier-Journal*, June 20, 1907, 6.
300. Bryant and Bryant, *Lawrenceburg*, 26.
301. "Funeral Services for Davis Brown, 50, Set for Wednesday," *Louisville Courier-Journal*, June 14, 1932, 3.
302. "Brown Interests Plan to Resume Distilling," *Louisville Courier-Journal*, September 26, 1933, 2.
303. Bryant and Bryant, *Lawrenceburg*, 26.
304. Ibid., 35.
305. "J.T.S. Brown Jr. Dies at Home in Florida," 8.
306. "Distillery Executive Brown Dies," *Louisville Courier-Journal*, September 10, 1946, 6.
307. "Capitalist, Ex-Resident, Hewett Brown Is Dead," *Louisville Courier-Journal*, November 8, 1952, 11.

Chapter 21

308. "Model License Founder Dies: George Garvin Brown, Distiller Succumbs Here," *Louisville Courier-Journal*, January 25, 1917, 1.
309. Ibid.
310. "Bourbon: The Lore of the Label," *Louisville Courier-Journal*, May 3, 2009, E2; "Civil War Surgeon Dies at His Home," *Louisville Courier-Journal*, August 16, 1909.
311. *Memoirs of the Lower Ohio Valley*, 283.
312. "From His Home: Funeral of Mr. Forman to Be Held Today," *Louisville Courier-Journal*, November 21, 1901, 10.
313. "Kentucky Whiskies," 18.
314. "From His Home," 10.
315. "Noted Distiller, Owsley Brown, Dies at 73," *Louisville Courier-Journal*, November 1, 1952, 1.
316. "The Bible and Prohibition," *Louisville Courier-Journal*, October 15, 1910, 5.
317. "Model License Founder Dies," 1.
318. Brown-Forman, "History."
319. Ibid.
320. "W.L. Lyons Brown Dies; Had Headed Distillery," *Louisville Courier-Journal*, January 6, 1973, 13.
321. Brown-Forman, "History."
322. "Brown-Forman Official, Owsley Brown, 73, Dies," *Louisville Courier-Journal*, November 1, 1952, 12.

323. Brown-Forman, "History."
324. Ibid.
325. "W.L. Lyons Brown Dies; Had Headed Distillery," 13.
326. Kyung M. Song, "Lee Brown Hands Top Post at Brown-Forman to Brother," *Louisville Courier-Journal*, July 23, 1993, 7.
327. Kleber, *Encyclopedia of Louisville*, 135.
328. Brown-Forman, "History."
329. Song, "Lee Brown Hands Top Post," 7.
330. Ibid.; Brown-Forman, "History."
331. Brown-Forman, "History."
332. Ibid.
333. Ibid.
334. Ibid.

Chapter 22

335. See Noe and Kokonis, *Beam, Straight Up*.
336. "Member of Distilling Family Dies, Cancer Claims Roy Beam Sr.," *Louisville Courier-Journal*, June 28, 1959, Section 1.
337. "Kentucky Whiskies," 18.
338. Pre-Prohibition Collector's Resource Site, "Early Times Distillery Co., Inc."; Cowdery, "157-Year History of Early Times Kentucky Whisky."
339. Scott and Scott, *Kentucky Bourbon Trail*, 118.
340. "Col. James B. Beam Dies at 83," *Louisville Courier-Journal*, December 28, 1947, Section 1.
341. Noe and Kokonis, *Beam, Straight Up*.
342. "Col. James B. Beam Dies at 83."
343. "T.J. Beam Dies: Retired President of Beam Distilling," *Louisville Courier-Journal*, May 3, 1977, D7.
344. Robert Schoenberger, "Frederick Booker Noe Jr. Retired Bourbon Distiller, Dies," *Louisville Courier-Journal*, February 24, 2004, B7.

Chapter 23

345. "Death Claims Big Distiller, Dietrich Meschendorf Succumbs in Texas," *Louisville Courier-Journal*, November 5, 1911, 3.
346. Ibid.

347. "Perpetually Enjoined, Judgements in Two Whiskey Trade Mark Cases," *Louisville Courier-Journal*, June 11, 1895, 8.
348. "Suit for Whiskey, Mill Creek Distillery Company Sues the Pleasure Ridge Park Company," *Louisville Courier-Journal*, June 23, 1897, 9.
349. "Old Kentucky Distillery Incorporated," *Louisville Courier-Journal*, July 6, 1901, 12.
350. "Some Other Purpose," *Louisville Courier-Journal*, December 14, 1899, 3.
351. "Warm Session: Expected When Distillers Association Meets," *Louisville Courier-Journal*, February 8, 1907, 5.
352. "Monarch Plant," *Louisville Courier-Journal*, September 4, 1904, A3.
353. "New Davies County Distillery Company, Organized and Incorporated," *Louisville Courier-Journal*, August 23, 1901, 5.
354. "Mrs. Meschendorf Dead," *Louisville Courier-Journal*, August 7, 1907, 10.
355. "Famous Waldeck Farm," *Louisville Courier-Journal*, June 17, 1908, 7.
356. "Death Claims Big Distiller," 3.
357. "Aids Church and Charity; Distillery Leaves Some $200,000 in Public Bequests," *Louisville Courier-Journal*, November 28, 1911, 2.
358. "Large Inheritance Tax on Meschendorf Estate," *Louisville Courier-Journal*, December 24, 1912, 2.
359. Ibid.
360. "Loss of $400,000: More than 12,000 Barrels of Whiskey Destroyed," *Louisville Courier-Journal*, November 18, 1911, 6.

Chapter 24

361. "Mrs. Thomas Dies: Succumbs After Illness of Five Weeks," *Louisville Courier-Journal*, October 27, 1905, 10.
362. Ibid.
363. "Full of Years, Maj. W.H. Thomas Succumbs to Infirmities," *Louisville Courier-Journal*, October 7, 1908, 12.
364. "Percy Thomas Dies," *Louisville Courier-Journal*, January 24, 1911, 3.
365. "Full of Years," 12.
366. "Thomas and Son Fail, Old Whiskey Firm Forced into Assignment," *Louisville Courier-Journal*, April 19, 1894, 8.
367. "$500,532.14: Liabilities of W.H. Thomas & Son, Go into Bankruptcy," *Louisville Courier-Journal*, April 19, 1901, 8.
368. "Full of Years," 12.

369. Young, *Pure Food*, 167.
370. "To Wind Up Affairs of W.H. Thomas & Son Company," *Louisville Courier-Journal*, April 7, 1907, 5.
371. "Full of Years," 12.
372. "Percy Thomas Dies," 3.
373. "Valuable Paintings Have Disappeared," *Louisville Courier-Journal*, April 29, 1929, 7.

Chapter 25

374. "To Be Buried Together: Double Funeral of Mr. and Mrs. David Sachs Will Take Place This Afternoon," *Louisville Courier-Journal*, June 8, 1898, 5.
375. "Kentucky Whiskies," 18.
376. "Fatal Stroke of Paralysis: Death of Mr. David Sachs, the Prominent Whiskey Merchant," *Louisville Courier-Journal*, June 7, 1898, 8.
377. Ibid.; "To Be Buried Together," 5.
378. "Sam Haas' Rites to Be Tomorrow," *Louisville Courier-Journal*, September 11, 1923, 20.
379. "Illness Is Fatal to Morris D. Sachs," *Louisville Courier-Journal*, November 17, 1932, 3.
380. "Morris D, Sachs Will Disposes of $150,000," *Louisville Courier-Journal*, November 26, 1932, 18.
381. "Edward Sachs Civic Leader, Dies at 77," *Louisville Courier-Journal*, September 22, 1938, 3.

BIBLIOGRAPHY

Books

Allison, Young. *The City of Louisville and a Glimpse of Kentucky*. Louisville, KY: Committee on Industrial and Commercial Improvement, Louisville Board of Trade, 1887.

Bernheim, Isaac. *Reflections of Isaac Wolfe Bernheim*. Louisville, KY: produced and published by Bernheim Arboretum and Research Forest, in partnership with Butler Books, 2016.

Bryant, William S., and Barbara S. Bryant. *Lawrenceburg*. Charleston, SC: Arcadia Publishing, 2012.

Campbell, Sally Van Winkle. *But Always Fine Bourbon: Pappy Van Winkle and the Story of Old Fitzgerald*. N.p.: Four Colour Print Group, 2012.

Haara, Brian. *Bourbon Justice: How Whiskey Laws Shaped America*. Lincoln: University of Nebraska Press, 2018.

Howell, Carl, and Don Waters. *Hardin and LaRue Counties: 1880–1930*. Charleston, SC: Arcadia Publishing, 1998.

Huckelbridge, Dane. *Bourbon: A History of the American Spirit*. New York: William Morrow, HarperCollins Publishers, 2014.

The Industries of Louisville and New Albany, Indiana. Louisville, KY: J.H. Elstner Publishing, 1886.

Johnston, J. Stoddard, ed. *Memorial History of Louisville from Its First Settlement to the Year 1896*. Vol. 1. Chicago: American Biographical Publishing Company, H.C. Copper Jr., 1896.

Kleber, John, ed. *The Encyclopedia of Louisville*. Lexington, KY: University of Lexington Press, 2011.

Louisville: Nineteen Hundred and Five. Louisville, KY: Commercial Club, 1905.

Memoirs of the Lower Ohio Valley: Personal and Genealogical. Vol. 1. Madison, WI: Federal Publishing Company, 1905.

Mitenbuler, Reid. *Bourbon Empire: The Past and Future of America's Whiskey*. New York: Penguin Books, 2015.

Noe, Fred, and Jim Kokonis. *Beam, Straight Up: The Bold Story of the First Family of Bourbon*. 1st ed. Hoboken, NJ: John Wiley & Sons, 2012.

Reigler, Susan. *Kentucky Bourbon Country: The Essential Travel Guide*. Lexington: University Press of Kentucky, 2016.

Scott, Berkley, and Jeannie Scott. *The Kentucky Bourbon Trail*. Charleston, SC: Arcadia Publishing, 2009.

Tamazzo, John C. *Bullets and Bourbon: True Stories of Whiskey, War, and Military Service*. Lincoln, NE: Potomac Books, 2018.

Young, James Harvey. *Pure Food: Securing the Federal Food and Drugs Act of 1906*. Princeton, NJ: Princeton University Press, 1989.

Magazines

Southwestern Reporter 9 (August 6, 1888–January 7, 1889). Containing all the current decisions of the Supreme Courts of Missouri, Arkansas, Tennessee and Texas; the Court of Appeals of Kentucky; and the Court of Appeals (Criminal Cases) of Texas.

Wine and Spirit Bulletin 17 (1903). Bulleting Publishing, Louisville and Cincinnati.

Newspapers

Cincinnati Enquirer.

Louisville Courier-Journal.

Web Resources

Brown-Forman. "History." www.brown-forman.com/about/history.

Bulleit Frontier Whiskey. "Stitzel Weller Distillery." www.bulleit.com/stitzel-weller-distillery.com.

Cowdery, Chuck. "The 157-Year History of Early Times Kentucky Whisky." The Whiskey Wash, August 3, 2017. https://thewhiskeywash.com/whiskey-styles/american-whiskey/157-year-history-early-times-kentucky-whisky.

Elijah Craig. www.elijahcraig.com.

Four Roses Bourbon. "Four Roses History and Heritage." https://fourrosesbourbon.com/heritage.

Heaven Hill Distillery. "Old Fitzgerald." www.heavenhilldistillery.com/old-fitzgerald.php.

Hotaling & Company. www.hotalingandco.com.

Old Rip Van Winkle. "Heritage." www.oldripvanwinkle.com/heritage.

The Pre-Prohibition Collector's Resource Site. "Early Times Distillery Co., Inc." www.pre-pro.com/midacore/view_vendor.php?vid=PAH11441.

Talbott, Tim. "Douglas Park Racetrack." Explore Kentucky History. https://explorekyhistory.ky.gov/items/show/320.

INDEX

A

B

C

D

E

F

G

H

I

J

T

U

V

W

ABOUT THE AUTHOR

Bryan Bush was born in 1966 in Louisville, Kentucky, and has been a native of that city ever since. He graduated with honors from Murray State University with a degree in history and psychology, and he received his master's degree from the University of Louisville in 2005. Bryan has always had a passion for history, especially the Civil War. He has been a member of many different Civil War historical preservation societies and round tables. He has consulted for movie companies and other authors; coordinated with other museums on displays of various museum articles and artifacts; written for magazines such as *Kentucky Civil War Magazine*, *North/South Trader*, *The Kentucky Civil War Bugle*, *The Kentucky Explorer* and *Back Home in Kentucky*; and worked for many different historical sites. In 1999, Bryan published his first work: *The Civil War Battles of the Western Theater*. Since then, Mr. Bush has had published more than fourteen books on the Civil War and Louisville history, including several titles for The History Press, including *Louisville During the Civil War: A History and Guide*, *Louisville's Southern Exposition*, *Favorite Sons of Civil War Kentucky* and *The Men Who Built the City of Progress: Louisville During the Gilded Age*. Bryan Bush has been a Civil War reenactor for fifteen years, portraying an artillerist. For five years, Bryan was on the board of directors and curator for the Old

Bardstown Civil War Museum and Village: The Battles of the Western Theater Museum in Bardstown, Kentucky; was a board member for the Louisville Historical League; and was the official Civil War tour guide for Cave Hill Cemetery. In December 2019, Bryan Bush became the park manager for the Perryville State Historic Site.